Fish Anatomy, Physiology, and Nutrition

AQUARIOLOGY

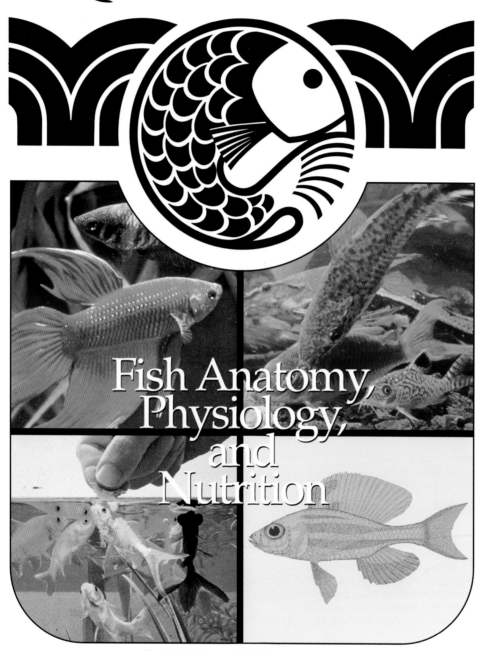

Fish Anatomy, Physiology, and Nutrition

Dr. John B. Gratzek, Dr. Howard E. Evans,
Dr. Robert E. Reinert, and Dr. Robert A. Winfree

Tetra⬤Press

Tetra Press
Aquariology: Fish Anatomy, Physiology, and Nutrition
A Tetra Press Publication

Gratzek, Dr. John B.,
Aquariology: fish anatomy, physiology, and nutrition; edited by
Dr. John B. Gratzek with Janice R. Matthews

L.C. Catalog card number 91-067992
ISBN number 1-56465-107-X
Tetra Press item number 16857

1. Gratzek, Dr. John B.   2. Matthews, Janice R.

Tetra would like to gratefully acknowledge the following sources
of photographs and artwork:
For the chapter "Getting Started" © Dr. John B. Gratzek; for
"Anatomy" © Prof. Emer. Howard E. Evans; for "Fish Physiology"
© College of Veterinary Medicine, University of Georgia; for "Nutrition and Feeding" © Dr. Robert A. Winfree. Additional
illustrations were provided by Tetra Archives, except as otherwise noted.

Printed in Hong Kong.

First edition
10 9 8 7 6 5 4 3 2 1

Production services by Martin Cook Associates, Ltd., New York

Table of Contents

Foreword

The successful keeping and reproduction of ornamental fish in a confined environment is the foundation of the tropical, goldfish, and pond fish hobby which is enjoyed by millions of people throughout the world. Information, education, and knowledge for ornamental fish keepers to be successful is an obligation for those companies engaged in providing products for home aquariums and outdoor fish ponds.

Tetra is extremely pleased to have undertaken the educational Aquariology Series of books to enhance successful ornamental fish keeping. *Fish Anatomy, Physiology, and Nutrition* is an excerpt from the Aquariology Master volume, *The Science of Fish Health Management*. To understand and appreciate fish, there is no more informative book than *Fish Anatomy, Physiology, and Nutrition*.

The Aquariology Series should be the basis for any library of ornamental fish keeping books. Here is a complete listing for your reference:

The Science of Fish Health Management
Item #16855 ISBN # 1-56465-105-3

Fish Breeding and Genetics
Item #16856 ISBN # 1-56465-106-1

Fish Anatomy, Physiology, and Nutrition
Item #16857 ISBN # 1-56465-107-X

Fish Diseases and Water Chemistry
Item #16858 ISBN # 1-56465-108-8

Alan R. Mintz
General Manager
Tetra Sales (U.S.A.)

Getting Started with Aquaria

John B. Gratzek

The simplest definition of an aquarium is a container capable of holding water in which fish and other aquatic organisms can live over a long period of time. Aquariology is the study of keeping fish in aquaria. It explores the reasons for keeping fish and requires a knowledge of the biological characteristics of fish, as well as of such aspects of their husbandry as feeding and nutrition, reproduction, water quality management, sources and control of stress, and disease control.

This book is written with the hope that it will provide both basic knowledge for the beginning aquarist and more specific information for advanced aquarists, producers of ornamental fish, and those wishing to keep fish as laboratory research subjects.

Choosing Appropriate Equipment

Selecting the aquarium: Obvious considerations will determine the size of the aquarium. Available space in a particular room, the size and type of fish you wish to keep, and the cost are factors which will influence your decision. In general, larger aquaria for a given number of fish will result in fewer problems than smaller aquaria with the same amount of fish. Pollutants accumulate more slowly in a larger tank, increasing the interval between required water changes. A larger aquarium also has the advantage of flexibility—providing space for several species of smaller fish or one or two very large fish. Additionally, a larger aquarium will provide the space for planting an interesting variety of living plants.

For the beginning aquarist, it is recommended that the aquarium be no smaller than 10 gallons (38 liters) and preferably larger. A 30-gallon (114-liter) aquarium would be an ideal size for the novice. If lightly stocked, a tank of this size provides sufficient volume of water to dilute out accumulated fish wastes. At the same time, it affords ample room for fish growth and addition of new residents. For breeding purposes, a larger aquarium will provide space for particularly territorial fish as well as escape room for small fish.

Wholesalers in the pet-fish trade prefer to use aquaria between 25 and 30 gallons (95 to 114 liters) in size. To facilitate netting fish, as well as cleaning, tank height is minimized. Retailers usually employ a series of 10- to 15-gallon (38- to 57-liter) aquaria to display freshwater fish; larger aquaria house saltwater fish and the larger freshwater species.

Aquaria are sold in many shapes, from rectangular to hexagonal or cylindrical. If due attention is paid to water quality, fish will do fine in any of them. Very deep aquaria may be difficult to clean; additionally, plants may not do well in very deep aquaria because of poor light penetration to the aquarium bottom.

A sturdy stand must be selected and carefully balanced to prevent the weight from shifting once the aquarium is filled with water.

It is a good idea to clean a new aquarium to remove fingerprints and to make its glass surfaces as clear as possible before filling it with water. A small amount of dishwashing detergent dissolved in lukewarm water is perfectly acceptable for cleaning aquarium glass, provided that the aquarium is *thoroughly* rinsed with warm water afterward to remove all traces of cleanser. Scale which has collected at the water line of used aquaria can be removed with commercially available preparations designed for the removal of lime deposits. Weak acid solutions such as vinegar may also be used to remove lime deposits. Some soaking time is necessary for removal of scale regardless of what preparation is used. An alternative method is to scrape scale away with a razor blade.

Supporting the aquarium: The stated capacity of an aquarium is somewhat more than the actual volume. For instance, a standard 10-gallon (38-liter) aquarium with inside measurements of 19.25 inches (38.9 centimeters) long, 10 inches (25.4 centimeters) deep, and 11 inches (27.9 centimeters) wide will hold about 9.16 gallons (34.7 liters). Addition of gravel, rocks, or inside filters will further reduce the water volume.

One gallon of water weighs 8.34 pounds; consequently the water in a "10-gallon" aquarium—about 9.16 gallons—would weigh about 76 pounds. (The water in a "38-liter" aquarium would weigh about 35 kilograms.) Gravel adds still more

weight. If a filled aquarium is either moved or placed on a stand which is not level, the resulting twisting stress on the glass is likely to result in breakage and leaks.

Aquarium stands constructed from metal or wood are available in a wide range of styles. Stands which include a cabinet below the aquarium have the advantage of hiding equipment such as air pumps and canister filters, and can be used for storage. For research purposes where many tanks are used, double-tiered racks constructed of 2-by-4-inch lumber (standard construction studs) bolted together can be easily constructed. A rack need not have a complete solid surface to support the aquaria; placing either end of the aquaria on a 2-by-2 or 2-by-4-inch board is adequate. Racks can be built so that the long axes of the aquaria are side by side. This configuration allows an investigator to utilize space more efficiently.

Providing cover and lighting: Most aquaria are manufactured so that various types of covers or hoods fit them snugly. If a cover is not used, evaporation can be a problem. As water evaporates, the concentration of dissolved minerals and organics (uneaten food, plant detritus, fish wastes, etc.) will tend to rise. A cover will keep the air directly over the water closer to the temperature of the water, minimizing heat loss and maximizing the efficiency of the heater. The cover also prevents fish from jumping out of the aquarium, an otherwise very frequent occurrence.

Most aquarium hoods have spaces provided for the installation of artificial lighting. Either ex-

For research purposes, an arrangement such as this one at the University of Georgia's College of Veterinary Medicine works well.

There are many lighting, filtration, and heating systems available on the market. Fluorescent light fixtures are common and varied.

clusively incandescent or fluorescent illumination can be used. In some specially designed units, a combination of both types is provided. Incandescent fixtures are less expensive than fluorescent ones, but incandescent bulbs will use more electricity and will give off more heat than fluorescent bulbs, which burn "cooler" and will not substantially affect the water temperature. Another distinct advantage of fluorescent lights is that they can be purchased in sizes which cover the entire length of the aquarium and consequently provide an even distribution of light for plant growth.

The aquarist has a wide choice of types of fluorescent bulbs which emit various spectra of light waves, some of which will stimulate plant life, including algae. If living plants are not used in the aquarium, bulbs are available which are more suitable for highlighting the coloration of a tank's fish than for stimulating plant growth. Full-spectrum bulbs for plant growth are available under a variety of trade names. The wattage required for optimal plant growth depends on the size of the

aquarium. In general, 1.5 watts per gallon of water is adequate. For example, a 30-gallon (114-liter) aquarium would require a single 40-watt tube or a pair of 20-watt tubes. The performance of fluorescent tubes will degrade over time. Their replacement as often as every six months is recommended by some experts to assure optimal conditions for plant growth.

For research purposes, individual aquarium lighting is not required unless the experimental protocol requires a definite photoperiod. Generally, outside rooms with windows will provide enough natural light to facilitate observing fish. In rooms where natural light is unavailable, establishing a photoperiod of between eight and twelve hours using normal room lighting is advisable.

Buying a heater: A desired water temperature can be easily maintained by using a thermostatically controlled immersible water heater. Without heaters, water temperature will fluctuate with room temperature.

Cold-water fish such as goldfish prosper over a wide temperature range. For example, goldfish will overwinter in iced-over backyard ponds. Most species of tropical freshwater fish as well as marine tropical fish do well at 75 degrees F (24 degrees C), but can tolerate temperatures 10 degrees F (6 degrees C) above or below that optimum at which the metabolic processes of the fish are at maximum efficiency. However, as temperatures drop lower than an acceptable range for a tropical variety, there is a depression of all body functions, including appetite, growth, and the immune system. Temperatures above the optimum range can cause stress by reducing the amount of available oxygen in the water, resulting in increased respiration. Increased respiratory rates can accelerate the development of disease problems relating to poor water quality or to the presence of parasites which affect gills. For special needs, where very cold water is necessary, refrigerated aquaria are available.

The wattage of a heater describes its power— the more wattage, the more heat will be delivered to the tank. Larger aquaria will require heaters of higher wattage, as will aquaria which are situated in cooler areas. As a general rule, use from 3 to 5 watts per gallon of water, depending on the room temperature. For example, if the room temperature is kept 8 to 10 degrees F (4 to 6 degrees C) lower than the desired aquarium temperature, as in a basement area, use a heater delivering at

Proper aeration of aquarium water will ensure the success of plant growth in the "natural aquarium."

least 3 watts per gallon. Buy an accurate thermometer to use when setting the heater's thermostats and for periodic checks of water temperature. Some brands of aquarium heaters have their thermostats set to regulate water at a specified temperature.

Both immersible and fully submersible heaters are available, as are low-wattage heating pads that are placed outside and under the tank being heated.

Buying an air pump: An air pump is a necessity for a modern aquarium and serves many functions. A simple diffuser stone positioned at the bottom of an aquarium serves to circulate water by the upward movement of air bubbles. As bubbles contact the surface of the water, the agitation increases the air–water interface, causing an increase in the rate of diffusion of atmospheric gases into the water and of dissolved carbon dioxide from the water. Air pumps can be used to move water through gravel beds, in conjunction with outside filters, and through cartridges containing water-softening or ammonia-removing resins. The "lifting" of water is accomplished by directing a stream of air bubbles through a tube. The upward buoyancy of the bubbles acts like a piston moving water through a tube. A variety of air pumps is available, and all generate some sound which should be evaluated prior to purchasing the pump.

Filters and filtration materials: All of the many varieties of filters available for use in aquaria can be categorized functionally as mechanical, biological, or chemical. Many combine two or more of these modalities in a single unit.

A mechanical filter functions by trapping suspended particulate matters which could include uneaten food, fish wastes, or any kind of biological or inert particles, in a filter matrix. The size of particle which a mechanical filter will remove and the time required for removal depend on the density of the filter material. Filter media include gravel, floss, foam, or inert particulate materials such as diatomaceous earth. These act as a mechanical barrier to fine suspended particles when adsorbed to a filter screen. Mechanical filters will eventually clog and their media will require cleaning or replacement. The time required for clogging is related to pore size. Filters with a pore size small enough to retain bacteria, for example, if installed in an aquarium without some sort of prefilter, would last a matter of minutes prior to clogging. In aquaria, mechanical filters are expected to remove large particles. Removal of particles as small as the free-swimming stages of most protozoan parasites by filters is possible. However, the pore size of the medium must be small enough to trap the parasites and there must be no possibility for the parasites to bypass the filter as the medium clogs. Some parasites have the ability both to swim and to change shape. This enables them to pass through filter materials, much like a water-filled balloon being forced through a small opening.

Biological filters oxidize fish waste products, primarily by changing ammonia to nitrates. The bacteria involved in this process, collectively known as nitrifiers, are common in nature and

Scanning electron micrographs (3000x) show one of the major effects of conditioning. Unconditioned gravel (top) is barren of life, but after conditioning (bottom), bacteria are visible on the gravel surface.

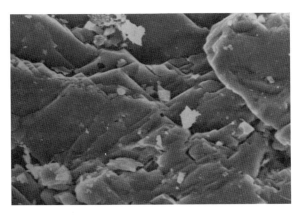

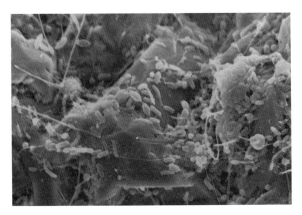

are introduced into the aquarium along with water and fish. They also are called chemo-autotrophic bacteria, because they require ammonia and nitrite ions for their growth. Bacteria of the genus *Nitrosomonas* utilize ammonia excreted by the fish as an energy source and oxidize it to nitrite ion. A second group of bacteria, belonging to the genus *Nitrobacter,* oxidizes nitrites to nitrate ion. These nitrifying bacteria initiate the conversion of nitrogenous wastes to free nitrogen. The second stage of the process, denitrification, is carried out by a different set of bacteria in the absence of oxygen. This makes it impractical to incorporate denitrification into home aquarium filter systems. Nitrifiers gradually colonize the surface of gravel, floss, foam filters, tubing, and any other solid surface, including the inner surface of the aquarium glass. (Note the scanning electron micrographs showing nitrifying bacteria on the surface of aquarium gravel.)

Chemical filtration entails passing aquarium water through some substance capable of changing the chemistry of the water. The type of change produced will depend on the substance included in the filter. Common chemically active filter media include:

1. Activated carbon. The physical structure of activated carbon includes a network of spaces responsible for adsorptive capacity. Activated carbon will adsorb a wide variety of organic substances, including color- and odor-producing substances. It effectively removes from solution dyes and chemicals used for treatment of fish disease problems, as well as dissolved heavy metals such as copper. It will remove neither ammonia nor nitrite ion from solution, nor will it soften water. Its primary use in home aquarium systems is to clarify water. Many manufacturers supply disposable inserts such as floss pads permeated with carbon particles or bags of activated carbon. Periodic replacement is necessary since temperatures required for reactivation of the carbon approach those attained in a blast furnace.

2. Ammonia-adsorbing clays. Also known as zeolites, these clays are sold in the form of chips. They require rinsing under a running tap prior to use in order to avoid clouding the aquarium's water. Many have the capacity to adsorb positively charged cations such as ammonium (NH_4^+) and can be used in filters. Since some zeolitic clays will also remove other types of cations such as calcium or magne-

sium, they also act as water softeners.

3. Ion-exchange resins. In some areas, water is "hard"; that is, it contains extremely high levels of calcium and magnesium ions. Frequently, the pH of such water is relatively high (7.8 to 9.0). Although a surprising number of fish can tolerate high levels of these minerals in water, many species will only breed under softer, more acidic water conditions. Thus many fish culturists prefer to adjust pH downward in their tanks. Doing so is difficult in the presence of calcium carbonates because the latter have a buffering effect. However, synthetic resins can be placed in a filter to soften water.

Resins which exchange sodium ions for calcium and magnesium ions are called cationic exchangers. When water is passed through this type of resin, a water test will indicate that the water has been softened. Many aquarists utilize softened water without problems for the fish. If softened water is used, the addition of a few grams of magnesium salts (Epsom salts) and calcium salts in the form of dolomitic limestone and/or oyster shell may be indicated.

The use of "mixed-bed" resins in a filter will essentially remove both cations (calcium and magnesium) and anions such as sulfates and carbonates. The resulting water is then said to be deionized. Fish cannot tolerate completely deionized water. However, partial deionization may be necessary in lowering pH in some hard-water areas.

4. Oyster shell or coral gravel. These media are usually used in a filter in areas where soft water has a tendency to become acidic abruptly. These materials contribute calcium carbonate to the

Attached to the side of the aquarium, this outside power filter has disposable inserts to facilitate the periodic cleaning that all such filters require.

water, increasing hardness and buffering capacity. In soft-water areas of the country, water in an unbuffered aquarium may decrease in pH to a point where fish are severely stressed or die.

5. Peat moss. Peat moss has been used in filters to soften water, usually for breeding purposes. It is likely that peat moss releases a hormone-stimulating substance into solution which induces spawning. Use of peat moss in a filter will impart a light brown color to water.

Choosing a filter: There is no good or bad filter. The various types available have distinct applications, depending on a variety of factors, including expense, tank size, number and/or size of fish kept in an aquarium, and whether the aquarium houses saltwater or freshwater fish. Practically every hobbyist, experienced retailer, or authority will have his or her own strong opinions on exactly what is best, but successful filtration always has both a mechanical and biological component. (Chemical filtration is required on a basis of need for special water requirements.)

These processes can be carried out using very simple or very expensive filter units—fish do not know the difference as long as the water quality is good. All filters provide for the movement of water through the filtering material, either by the air-lift principle or by electrically driven pumps. Filters may be located inside or outside of aquaria. All eventually tend to clog, resulting in reduced flow rates and inefficient filtration. All filter media, whether floss, foam pads, activated carbon, gravel, or plastic rings used for

A corner filter, which works by the air-lift principle, is appropriate for a small aquarium.

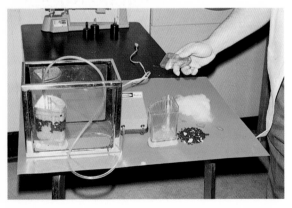

mechanical filtration, will eventually be colonized by nitrifying bacteria. The bacteria are firmly attached to the filter materials and are not removed by vigorous rinsing. Naturally, hot water, soaps, and various disinfectants will kill these bacteria and destroy the beneficial effects of biological filtration. All filters require periodic cleaning to remove debris which, although trapped within a filter matrix, is in fact still adding to the organic pollution of the aquarium water.

Since the mechanical and biological functions of filtration materials are so intertwined, it is recommended that when a filter is serviced, at least some of the media contained therein never be discarded. The easiest way to accomplish this is to include gravel, plastic, or ceramic rings in a filter along with disposable filter media such as activated carbon. Since activated carbon loses its filtering capacity after a period of time, placing it in a bag within the filter will simplify changing.

1. Corner filters. Included in many aquarium "beginner kits," an inexpensive corner box filter can be effectively utilized in smaller aquaria. Most corner filters are operated by the air-lift principle. Their filtration capacity is limited. For general filtration purposes, they function best when a small permanent sack of gravel or similar substrate is incorporated along with mechanical and chemical filter material (floss and activated carbon) to ensure that some bacteria-laden "conditioned" material remains after cleaning. Corner filters are frequently used in small aquaria for holding fish during a quarantine period or during a brief treatment period. Many aquarists interested in breeding fish use corner filters in their spawning setups.

2. Outside power filters. Most outside filters are constructed so that they can be easily hung from the rim of an aquarium. Various types of electrically driven filters are available, but most are driven by rotary impeller motors. However, an outside filter is defined only by location, and air-lift driven units are also available.

Outside power filters can be loaded with any type of filter material that meets the aquarist's needs. These generally include floss or foam pads, positioned to keep larger particulate matter from clogging activated carbon or other filter material. If gravel or other materials such as ceramic or plastic rings are included, the filter will in time develop a biological function. The waste-processing capacity of such units will be limited when

Outside power filters can be set up easily, and trapped debris rinsed periodically, with little disturbance of the enclosed environment. Foam (sponge) filters are also popular in situations where no gravel bed is to be used in the aquarium.

compared to an undergravel filter. However, aquarists wanting to avoid gravel beds for any reason will find such modified outside filters a useful alternative for use in breeding, fry rearing, or quarantine tanks.

All outside filters require periodic cleaning. Obviously, debris trapped in a filter remains in contact with the aquarium water. Depending upon the flow characteristics of the brand of outside filter being used, water may bypass the filter media as the filter clogs. This will result in less efficient filtration with little or no change in flow rate. By comparison, other filter types do not allow bypassing of water, so flow rate slows down as the filter clogs.

Cleaning outside filters is much easier if particulate media such as gravel, ceramic rings, or activated carbon are placed in separate net bags. Floss or foam pads should be cleaned whenever debris buildup is evident. Gravel bags need only

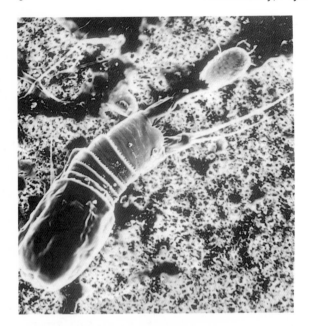

A thriving community of invertebrates will eventually develop within the filter system. This electron micrograph shows rotifers living upon a foam filter (260x).

tanks occasionally require isolation for treatment purposes, for maintaining biological filtration in quarantine tanks, and in tanks used for breeding fish.

Foam filters which have been in use for some time and which have developed a bacterial flora can be used to maintain and "seed" new aquarium systems with nitrifying bacteria. This is a simple way of avoiding the accumulation of ammonia and nitrites in freshly set-up aquaria that characterizes "new-tank syndrome."

Like all filters, foam filters will develop a thriving community of various invertebrates which provide food for fish in the aquarium. Many of these invertebrate forms are rotifers of various types. Rinsing foam filters under a stream of tepid chlorinated water will not kill the bacteria present, but may reduce the population of invertebrates temporarily. However, soaps, disinfectants, or deter-

Undergravel filters, the author's preference, come in many shapes and sizes. The simplest is a small flat plate (top) powered by a single air lift. For a larger aquarium, an undergravel filter with larger lift stacks powered by several air lifts (below) is ideal.

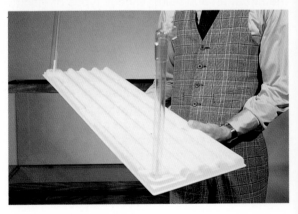

be rinsed under tepid tap water. Activated carbon should be replaced according to the manufacturer's suggestions.

3. Foam filters. Foam pads function very efficiently as mechanical filter media. They can be used in place of floss or floss pads in any type of filter. They eventually will develop a flora of nitrifying bacteria and function well as biological filters. Foam blocks must be rinsed periodically to restore their mechanical function. The most popular types of foam filters are powered by the air-lift principle. Larger units driven by rotary-impeller power heads are also available. Foam filters certainly are less efficient than outside power filters in their mechanical action, but will ultimately develop a good biological function as well. They are useful in systems where gravel is not required or desired as a biological filter bed. They also can be used either as a tank's sole filter or as a supplement to other kinds of filters. Single pads are suitable for smaller aquaria; multiple filters can be used in larger systems. Such filters are employed with considerable success for research aquaria at the University of Georgia for a variety of experimental uses. They are especially useful as adjunct filters in recirculating systems where

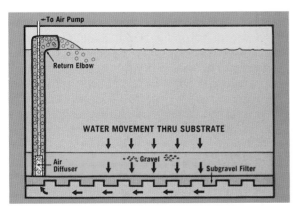

As water moves through the gravel into the undergravel filter plate, it carries oxygen to the nitrifying bacteria in the gravel. These bacteria process the fish wastes, converting ammonia to nitrites and eventually to nitrates. In this system, the entire gravel bed essentially acts as a biological filter.

gents may kill or inactivate all of the bacterial and invertebrate populations. Similarly, drying foam filters will inactivate bacteria and invertebrates. However, filters can be stored wet for several days without appreciably reducing their biological effectiveness.

4. Undergravel filters. Undergravel filters consist of a plastic plate equipped with one or more air lifts. At least 3 inches (7.6 centimeters) of washed gravel should be put over the plate. The air flow displaces water through the lift stacks, circulating water through the gravel bed. This brings both oxygen and organic wastes into contact with bacteria attached to the gravel. As long as this flow is maintained, the entire gravel bed of such a filter is biologically active. In a gravel-bottomed aquarium without water circulation, only the top centimeter or two is biologically active. Undergravel filter plates are sold in sizes adequate for the largest aquaria or for goldfish bowls. The flat plate types appear to be adequate for smaller aquaria. However, for larger aquaria, models with corrugated bottoms and larger-diameter lift stacks are likely to produce better water circulation. Undergravel units are widely employed in marine aquaria, where dolomitic limestone or coral gravel is used as the filter bed. As ammonia or nitrite poisoning is of special concern in marine tanks, the rapidity with which an undergravel filter removes these toxic substances makes it a natural choice for the marine tank.

At this writing, it appears that more and more aquarists are utilizing undergravel filtration for freshwater aquaria. Because undergravel filters act as both mechanical and biological filters, they need not be supplemented with other types of filters if (and this is an important "if"!) debris is periodically removed from the filter bed and water is changed regularly. The gravel bed is easily cleaned when water is changed by using a distended siphon tube to remove debris deep in the gravel bed. If water is changed on a regular basis, water can be kept clear without the use of activated carbon.

Of course, additional filters of any type can also be used with an undergravel filtration system. For example, an external power filter loaded with a filtering substance such as activated carbon, peat, or ammonia-adsorbing clays could be used to effect some desired change in water chemistry. Although undergravel filters are usually powered by air displacement, an alternate method is to place rotary impeller–driven units known as power heads on top of the lift stack. Placing a siphon tube from an outside power unit in a lift stack is a good way to operate an undergravel filter in conjunction with an outside power filter.

One disadvantage of undergravel filters is that some fish burrow into gravel or else actively move gravel about the bottom. This exposes the filter plate, creating breaks in the gravel which lead to a reduced water flow through the bed. This can be avoided by placing a plastic screen 3 to 4 centimeters (about 1.5 inches) below the surface of the gravel. Perhaps the chief disadvantage of undergravel filters is their immobility. There is no way to remove an undergravel filter from a tank that is being treated with therapeutic agents toxic to nitrifying bacteria, such as methylene blue, formaldehyde, or many antibiotics. Another frequently cited disadvantage is that rock formations reduce the effective surface available for water flow. From experience, I do not feel that rock formations cause sufficient blockage of the filtration surface to be of any consequence.

There is some controversy regarding the suitability of an undergravel filter for plant growth. It has been suggested that root movement which may be associated with the use of undergravel filters inhibits plant growth. However, experts in the field of hydroponics are able to grow a wide variety of plants without any root substrate and suggest that the aeration and micronutrients supplied to roots by an undergravel filter would bene-

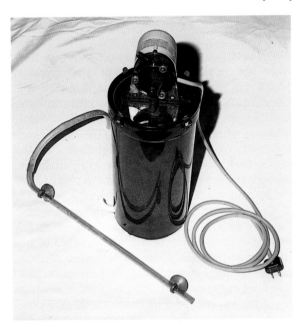

Canister filters are suitable for larger aquaria and can be loaded with a variety of filtering materials such as ceramic rings, ammonia absorbers, floss, and activated carbon.

fit the plants. Some recommend that when plants are to be used with an undergravel filter, the gravel layer should be increased to approximately 5 inches (13 centimeters). An alternative is to purchase plants which have been propagated in plant plugs containing fertilizers in a root-support growth medium. Plants can also be placed in small pots with good potting soil. It is important to place a layer of aquarium gravel over the soil in pots to stabilize it under water.

5. Canister filters. Canister filters take their name from their general shape. Their powerful motors pull water through a sealed container filled with various filter media. These filters are particularly useful in large or heavily stocked aquaria, which require a greater filtration capacity. Canister filters can be purchased in a variety of sizes. These units have two distinct advantages over other kinds of filters: they have enough volume to accommodate a series of filter substrates stacked in series, which greatly enhances their effectiveness, and they can be placed in a location remote from the aquarium. This latter feature is useful in display aquaria, for filters can be located in an adjacent work area.

Newer designs of canister filters are available in which the impeller is located at the bottom of the canister. This feature makes it very easy to prime the units and ensures that the impeller assembly never runs dry. Because of the increased size of the motor when compared to that of outside filters, some canister filters can be used in conjunction with inserts which will support a film of diatomaceous earth. These units can be used for water "polishing."

The disadvantage to canister filters is their higher cost in comparison to other types of filters. Another is that since outside canisters require tubing running to and from the aquarium, connectors must be carefully tightened and rechecked periodically lest leaks develop.

Buying gravel and ornaments: The aquaria in homes or for display purposes require a gravel base which, from an aesthetic viewpoint, mimics the bottom of a pond and provides a good base for rooting plants. For most freshwater aquaria, it is important to use quartzite or granite gravel, which will not contribute carbonate ions to water. There are advantages to using calcareous substrata in tanks housing fish that prefer hard alkaline water, such as African Rift Lake cichlids or most live-bearers.

The size range of the gravel particles should be approximately 4 to 6 millimeters (3/16th of an inch plus or minus 1/16th). The particle size is important for several reasons. If an undergravel filter plate is to be used, the spaces between gravel particles will allow free water flow and ample aeration for bacteria which will eventually colonize the surface of the gravel particles. The depth of the gravel bed will depend on whether or not an undergravel filter is used and whether live plants are desired.

Sand is not recommended as an aquarium substratum. The extremely small particle size results in packing and reduces water flow. Waste that breaks down in the resulting anaerobic conditions will generate hydrogen sulfide and other highly toxic substances. Aquatic plants also require a bed which will allow diffusion of nutrients to roots.

Marine aquaria usually are equipped with undergravel filters. Calcareous gravels, which contain carbonates, are recommended for such tanks. These include dolomitic limestone, crushed oyster shell, and coral gravel, materials containing high levels of carbonates. The slow release of carbonates in marine aquaria tends to

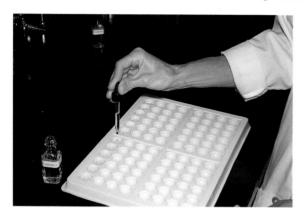

A multi-well tray allows many water samples to be tested at once.

buffer water towards the desired high (7.8 to 8.3) pH range.

Regardless of type, gravel intended for aquarium use requires removal of pulverized particles which can cloud tank water. Gravel should be rinsed under a tap while being stirred briskly, until the water runs clear.

Colored aquarium gravel is available. Although selection of color is largely a matter of human taste, white gravel reflects more light and may stress those species of fish which prefer dark areas in an aquarium.

Decorating an aquarium provides benefits both for the aquarist and for the fish. Many fish are territorial, and rock formations, plants, and a variety of other decorations will provide needed territorial landmarks and boundaries. Also, smaller fish may need to escape from larger species by taking refuge in small nooks provided by rock formations. A decorated tank also provides shade for those fish which prefer darker areas.

Not all objects are equally suitable for aquarium decoration. Coral, seashells, limestone, and marble will dissolve in fresh water and may increase the pH to an unacceptable level. Copper objects, galvanized metals, or steel can cause heavy metal poisoning, especially in areas where the water is soft and the pH is on the acid side of neutral. Rocks, driftwood, or gravel taken from streams or ponds should be soaked in a disinfectant such as chlorine bleach, then rinsed well, to avoid the introduction of snails and other unwanted invertebrates such as planarians and free-living nematodes.

The use of a background behind the aquarium serves to beautify the tank as well as to create the darker area preferred by certain shy species of fish. A variety of selections is available for outside use, including paints, paper with fresh- or salt-water motifs, and plastic materials constructed to create a three-dimensional illusion. Inside backgrounds constructed of a variety of waterproof inert materials are also available.

Buying water test kits: The serious aquarist should invest in test kits which will enable him or her to measure pH, hardness, ammonia, and nitrite levels. For saltwater aquaria, a hydrometer and copper test kit are also recommended. Most test kits sold for application in aquaculture are easy to use. They are based on color changes in the sample being tested, which is then compared to a color standard. Some kits are supplied with liquid reagents which over time may deteriorate. Others provide powdered reagents, which may be more stable over time. If it is necessary to test many aquaria, these kits can be used in conjunction with a tray containing many wells, each containing a water sample from a tank to be tested. This method provides a quick overview of water conditions in many aquaria.

Setting Up the Aquarium

Choosing a location: We will assume that you have purchased or built a stand which will support the weight of the fully set-up aquarium. If the tank is not level or appears to be unstable, it may be a good idea to shim the stand, or, alternatively, to fasten it to a wall to avoid accidental tipping. This can be done using an L-shaped piece of metal. The aquarium should be placed in an area where it is likely to be viewed, but not in an area where accessibility is limited. Regular maintenance will be much easier if it is possible to have working room above and behind the aquarium. Do not place the tank in direct sunlight. Otherwise, algae will rapidly accumulate and the aquarium may overheat. However, indirect light or even a short period of direct light in addition to the use of overhead lighting can be useful in stimulating plant growth. Placing aquaria close to air conditioning vents or over heat vents can complicate the task of regulating water temperature.

Equipping the interior: Once the aquarium is in the desired position, install the undergravel filter plate if this type of filter is going to be used.

Be careful not to displace gravel substrate while filling the tank with water. Rocks and plants can be more easily placed in tanks filled one-third to one-half capacity with water.

Then add washed gravel to a depth of 2 to 3 inches (about 5 to 8 centimeters) if an undergravel filter will not be used, to a depth of 3 inches (7.6 centimeters) if an undergravel filter will be used without live plants, and to a depth of 5 inches (12.7 centimeters) if such a unit will be used with

plants. For better plant growth, mix a soil additive or a proprietary slow-release fertilizer with the gravel. These products are available at pet-supply outlets.

After the washed gravel is added, fill the aquarium with water to about one-third of capacity; rock formations and plants are easier to set in place if some water is present. If an undergravel filter is being used, direct water over a shallow pan to avoid displacing gravel under it. Some aquarists prefer to slope the gravel slightly toward the front of the aquarium, which they claim facilitates removing debris from the aquarium.

Construction of caves and recesses makes for more interesting viewing while providing more timid fish with shelter. Do not use any type of rock which has the potential for releasing minerals, such as limestone, marble, or clays. Rocks collected from streams can be used. However, they should be thoroughly cleaned by brushing with water, rinsed, and dried prior to placement in the aquarium.

Plants, be they living or plastic, should be positioned with the taller-growing varieties toward the back of the aquarium and in a position to hide lift stacks, siphon tubes, or heaters which you will be adding. Because living plants require a favorable water quality in addition to plant nutrients and good lighting, some experts suggest waiting to add plants until after the aquarium has been established for a period of time. Presumably, in an established aquarium, nitrification would be in place and nitrates would be available for plant nutrients.

Although fish in ponds can tolerate water temperature fluctuations, in an aquarium there is no advantage to allowing such fluctuations, which can easily be avoided by installation of a heater. Heater placement will depend on the type purchased. Totally immersible types can be positioned horizontally at the level of the gravel, a location which may enhance plant growth. Models which are not totally immersible are usually clipped to the aquarium side; they require that the aquarium water level be kept at or above the level of the thermostat to avoid overheating of the water. With either type, promoting water circulation by positioning airstones at gravel level or using power filters for circulation will help ensure uniformity of water temperature throughout the aquarium.

Water temperature should be set prior to addi-

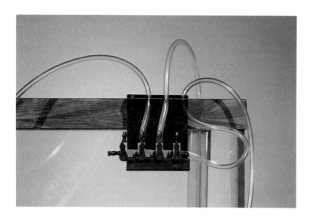

Loop the air line through notches in a tubing manifold to avoid the possibility of back-siphoning.

tion of fish to avoid any possible stress in the fish from drastic accidental changes in temperature. A good temperature for a wide variety of freshwater fish as well as plants is 75 degrees F (24 degrees C). Let the heater acclimate to the water for an hour prior to connecting the electricity. Then adjust it slowly in order to avoid possible overheating. A light will indicate whether the heating unit is on. For the first twenty-four hours, frequently check the temperature and adjust the thermostat as required to obtain the desired water temperature.

A few safety precautions should always be followed around tanks where a heater is in use. Never connect the heater unless the tube housing the heating element is immersed in water. Always disconnect the heater when changing water or lowering the tank's water level for any reason. Adjust the thermostat only when you have time to check the temperature of the water continually.

Providing aeration for the aquarium is the next matter of concern. Aquarium water is aerated by agitation of its surface by the outflow of power filters, a stream of bubbles produced by an airstone or by a stream of water splashed from the lift stacks of an undergravel filter. Gas exchange occurs at the surface of the water. Placement of an airstone toward the bottom of an aquarium will both circulate and aerate the water. Aeration of aquarium water provides oxygen required by fish, plants, and nitrifying bacteria. Aeration can always be increased by directing water flow from power filter outlets over the surface of the aquarium water. Undergravel filters have the built-in

advantage of "pulling" oxygenated water through the gravel bed.

When lines to airstones or lift stacks are attached to pumps located below the aquarium, there is a possibility of back-siphoning if the pump is accidentally disconnected or if there is a power outage. This can be avoided by positioning the air pump higher than the aquarium. If this is not possible, make sure that the air tubing to each outlet has a loop sufficiently high to avoid a siphon effect. Loops can be made by using the notches in a tubing manifold or by using plastic inserts in air tubing which prevent abrupt bending and the formation of anti-siphon loops. Check valves are also available, which can be placed in air lines to prevent back-siphoning problems.

A minimal agitation of the water surface from any type of aeration will generally result in oxygen levels of between 6 and 7 parts per million (ppm). Excessive agitation may disturb some more timid species such as discus fish. However, moderate circulation of water is tolerated by most fish and promotes plant growth.

At this point the aquarium should have gravel in place and be equipped with a filter, heater, and perhaps airstones. Fill the aquarium to the top with tap water. The air pump can be turned on immediately, but wait for a thirty- to sixty-minute period for the thermostat to adjust to ambient water temperature prior to plugging in the heater. The next step is to assure water quality that will support fish.

Dechlorinating the water: In many municipal water supplies, chlorine is added at the pumping plant to destroy bacteria pathogenic to humans. In tap water, dissolved chlorine concentrations usually measure between 0.2 ppm and 0.7 ppm, depending on the time of the year. Water for aquarium use must be chlorine-free, since even 0.2 ppm will kill fish by destroying gill tissues. Chlorine can be removed from tap water in three ways:

1. Aeration of water, resulting in diffusion of chlorine into the air. The use of a faucet-end aerator commonly found in households will aerate water and remove chlorine. Simply pouring water from one pail to another three or four times will also drive chlorine from solution. Letting tap water stand in pails for a few days also allows chlorine to dissipate gradually. This process can be speeded up by aerating the water with an air diffuser.

2. Passing water through activated carbon. Many of the faucet-end water purifiers sold to improve the taste of water are charged with activated carbon. Larger canisters available from water conditioning companies are also available for large-use situations such as would arise in a pet store.

3. Adding sodium thiosulfate to tap water immediately inactivates chlorine. Sodium thiosulfate is sold under a variety of trade names. One molecule of sodium thiosulfate will remove four molecules of chlorine. Based on this, 0.50 milliliters (10 drops) of a 1 percent solution of sodium thiosulfate would remove 0.5 ppm of chlorine from 10 gallons (38 liters) of city water. One drop per gallon of a 1 percent solution of sodium thiosulfate would provide a sufficient safety factor to avoid the consequences of fluctuations in chlorine levels.

Removing chloramines: In some municipal water plants, ammonia is added to react with chlorine to form chloramines, which then act as the disinfecting agent. The addition of sodium thiosulfate will neutralize both chlorine and chloramines. However, ammonia is released after the sodium thiosulfate combines with the chloramines, and this could be a problem to fish under conditions where there is little or no biological filtration.

In most home aquaria where biological filtration has been established, the routine use of sodium thiosulfate rids the water of chlorine; the remaining ammonia is quickly oxidized to harmless nitrates by the resident nitrifying bacteria. This assumes a very efficient biological filter and a relatively modest water change of no more than 25 percent of the tank volume at a time.

In newly established aquaria, or when most of the water has been changed at one time, sufficient nitrifying bacteria may not be present to oxidize ammonia. In these cases, sodium thiosulfate can be used in conjunction with ammonia-adsorbing media. The most readily available of these are certain clays (zeolites) sold under a variety of brand names. The ammonia-adsorbing chips should be placed in an outside or canister filter before making a water change to ensure that the ammonia released following dechloramination is quickly removed from the aquarium. An alternative method of neutralizing ammonia without using zeolites is simply to lower the pH of the water, if the tank's residents can tolerate it. At lower pH levels (6 to 7), the majority of total ammonia will be present as the nontoxic ionized form: ammonium (NH_4^+). In many areas of the country, pH can be lowered by adding buffers which are available in aquarium supply stores or by adding monobasic sodium phosphate, NaH_2PO_4.

Chloramines can also be removed from fresh water by the use of a high grade of unused activated carbon. Activated carbon which has been used may remove colors and odors, but will not remove chloramines.

In marine aquaria where pH levels are kept between 7.8 and 8.3, chloramine treatment with sodium thiosulfate would result in the generation of free ammonia. Zeolites are ineffective in removing ammonia from salt water, and lowering the pH in marine aquaria is not recommended. Before adding salt mix to tap water to make synthetic seawater, pretreat the requisite volume of tap water with sodium thiosulfate, then filter it through zeolite held in a household colander; alternatively, pass the tap water through virgin activated carbon prior to mixing it with salt.

If chloramines are present in locations where tap water is hard, with a high (7.8 to 9.0) pH, the ammonia resulting from treatment with sodium thiosulfate could injure fish if biological filtration has not been established. In such areas, pH may be difficult to adjust downward and water may have to be pretreated with sodium thiosulfate, then slowly poured through a pail containing zeolite chips before being used in the aquarium. (Holes in the bottom of the pail will facilitate this operation.) In any case, where chloramine removal is deemed necessary, treated water should be tested for total combined chlorine levels and ammonia to make sure that the treatment chosen was effective. Commercial products are also available that bind the ammonia produced by dechloramination into an organic complex that is harmless to fish. The complex is then metabolized by the biological filter.

Controlling pH: Water in various parts of the country may have different pH values, and in some cases the water may require some adjusting prior to addition of fish. Generally, fish can tolerate quite a wide pH range without problems. A pH of 6.5 to 7.8 for freshwater species is an acceptable range for maintenance of optimal health.

Purchasing and Adding Fish

Choice of fish: It has been customary to start new or unconditioned aquaria with hardy fish, supposedly more tolerant of ammonia and nitrites. There is no question that some varieties are less susceptible to nitrite intoxication than others. Our experience suggests that various species of tetras, such as the serpae tetra (*Hyphessobrycon callistus*), are less susceptible to nitrites than live-bearing fish such as swordtails, guppies, and platys. Common goldfish, zebra danios, and many barbs are also relatively hardy fish. If some form of conditioned filter—such as gravel from a conditioned aquarium or a conditioned foam filter—is added to a new aquarium, there will be less reason to fear the ammonia-nitrite problem.

A decision on what type of fish to eventually put into your aquarium is purely a matter of preference. However, you should be aware that some fish are incompatible with each other or with living plants. Different fishes also thrive in different water conditions. Some fish species may prefer brackish water. Others may do well in hard water with elevated pH, while still others may flourish in soft water with a lower pH.

A "community" tank exists when several species of fish are maintained together in an aquarium. In many cases, a few goldfish are included with species of live-bearing fish such as guppies or egg-laying species originating in South America. Such mixtures provide interesting visual variety but do not remotely reflect natural fish populations. If you are interested in goldfish, consider having an aquarium with nothing but goldfish. Alternatively, many aquarists like to create an aquarium with a few species of fish native to a particular part of the world. In many cases the schooling behavior of fish is not seen unless several fish of the same species are kept together.

How many fish in an aquarium?: In any new aquarium without an efficient biological filter system, just a few fish should be added initially. This introduction should be followed with regular water changes on a weekly basis for at least a month. As nitrifying bacteria develop in the filter material, more fish can be added.

The number of fish which an aquarium will support depends on several factors. A common rule which has been used by aquarists is that for every gallon (3.8 liters) of water, one may add 1 inch (2.5 centimeters) of length of freshwater fish or 0.5 inches (1.25 centimeters) of saltwater fish. Other aquarists suggest that the total inches or centimeters of fish which can be added should equal the number of inches or centimeters, respectively, which the aquarium measures along its long axis. A standard 10-gallon (38-liter) aquarium, for example, measures about 19 inches (48 centimeters) and thus could support nineteen 1-inch fish or twenty-four 2-centimeter fish.

It should be noted that some aquarists disregard all formulas and crowd their aquaria with fish. Their success is based on a good filtering system, a program of regular water changes, aeration, and due attention to nutrition and disease control. Nonetheless, it is generally better to have fewer fish in an aquarium to avoid deterioration of water quality and to minimize the risk of disease, which is enhanced by crowding.

Selecting healthy fish: Reputable retailers are not interested in selling an obviously sick fish, but often it is very difficult to detect fish which are carriers of a parasite and which with time will develop signs of disease. It is always prudent to select fish from aquaria where no disease has been evident over a period of time. Fish should be active, with a full underbelly. Signs of disease include clamped fins, lack of color, skin blemishes, white spots, excessive body slime, failure to eat, and inactivity. Some hobbyists will not purchase a fish (especially an expensive one) until they have observed it over a period of time in a retail shop. Particularly when evaluating marine fish, it is a good idea to ask a retailer if the animals in question have been routinely treated for parasites.

Quarantine: Quarantine refers to the isolation and observation of fish prior to introducing them into an aquarium. The objective of quarantine is to determine whether the specimen has a disease which could be transmitted to other fish. The assumption is that a serious disease is likely to develop during the isolation period. A quarantine period can vary in length, but fourteen days is common. In fish health management involving food fish, public aquaria, or fish used for research, fish may be routinely treated during the quarantine period. In cases where many fish are involved, a few fish may be killed and examined for parasites or chronic disease conditions. If disease is present, the fish are treated with a specific medication during the quarantine period. If a disease is detected which is either difficult or impossible to treat, a decision is made regarding the eventual use or disposition of the animals.

Quarantine of fish prior to their introduction into the home display aquarium is a rare, but nevertheless a recommended, practice. It is particularly advisable when adding new fish of questionable disease status to an established aquarium housing valued fish.

Adding fish to the aquarium: Although fish can live over a considerable range of water temperatures, sudden temperature changes can stress fish. It is a good practice to minimize stress by making sure that the temperature difference between the fish's transport container, usually a plastic bag, and the home aquarium is minimal (ideally less than 3 degrees F or 1.6 degrees C). In most situations, this can be done by floating the plastic transport bag in the aquarium water for ten to fifteen minutes. Keep the bag inflated during this period, since draping an uninflated bag over the side of an aquar-

Using a transport bag can allow you to introduce fish to the tank, while minimizing the shift in pH and temperature.

ium will minimize diffusion of oxygen by reducing water–air surface area.

Adding fish to the aquarium can be done in different ways, depending on the concentration of ammonia in the water of the transport container. Timely transport of a few fish in a plastic bag from a retail outlet to a home aquarium typically results in very low ammonia levels. However, on a commercial scale where hundreds of fish are transported, the ammonia levels in the transport water may be very high. Since the pH of the water in heavily packed bags is usually 6.5 to 6.8, the ammonia is in the nontoxic form. However, addition of fresh chlorinated water with a high pH (7.8 to 9.0) will serve to convert nontoxic ammonium to toxic ammonia, resulting in gill damage. This problem is particularly serious when marine fish are shipped considerable distances.

In most instances involving home aquaria, the practice of mixing aquarium water with the contents of the transport container, then adding the mixture to the aquarium, is unlikely to hurt the fish. Simply adjusting the pH of the aquarium water to approximately 6.8 to 7.0 will ensure that the ammonia levels remain low.

For wholesalers, retailers, and others who handle substantial numbers of fish crowded in bags, it is best to transfer the fish from the transport bag to the aquarium by the careful use of a net. The objective is to keep any polluted transport water and any associated disease organisms from entering the aquarium. Netted fish can be injured by contact and friction, especially if many fish are netted together. Netting injuries can be avoided by positioning a net just at the surface of a shallow container or pail filled with temperature- and pH-adjusted aquarium water. Fish in the transport bag are then "poured" into water but still contained by a net. A rapid transfer of fish into an aquarium can be effected with minimal contact of fish with the net.

Some wholesalers and brokers (trans-shippers) prefer to acclimatize fish by the slow addition of fresh water to the transport bag or to a container to which both fish and transport water have been transferred. In a matter of a few minutes, a total water change has been made and the fish can be transferred to aquaria without netting. This is an acceptable method provided that: (1) the pH of the incoming water does not differ from that of the transport water by more than 0.5 pH units in either direction and is not alkaline, (2) there is no great difference in water temperature, and (3) the water is dechlorinated.

It is always a good idea to determine the pH of the transport water, which generally will be between 6.5 and 6.8. Adjustment of aquarium water pH to between 6.8 and 7.0 prior to addition of freshwater fish is a sound and safe practice whenever the pH of the aquarium water is either much higher or much lower than the latter values.

Contrary to a widely held belief, most fish can tolerate rapid pH changes between the extremes of pH 6 and pH 9 if ammonia and other pollutants are not present. Fish in nature are often exposed to these variations without harm. Moreover, the author has experimentally shifted fish from pH 6 to pH 9 water without affecting their health. Nonetheless, overall it pays to err on the side of caution when contemplating pH changes in established aquaria. Both the direction and magnitude of pH changes must be evaluated, and this is a very complex subject. In addition, certain fish such as the neon tetra are quite intolerant of radical changes in water conditions. Thus, it is considered prudent not to alter pH by more than 0.5 units in a given twenty-four-hour period in a tank containing fish.

In the acclimation of marine fish, transport water should not be added to the aquarium, for the high pH of marine aquaria will ensure that toxic ammonia is present. After temperature equilibration, fish should be transferred to aquaria by careful netting or other appropriate methods, such as utilizing plastic containers with holes punched in the bottom as sieves. Fish can be caught easily, water allowed to drain, and the fish can be transferred quickly into the aquarium. As a rule, you can also avoid problems by refraining from adding transport water to aquaria.

Selected References

Baensch, H. 1983. *Marine Aquarists' Manual.* Tetra Press.

Hunnam, P.; Milne, A.; and Stebbing, P. 1982. *The Living Aquarium.* New York: Crescent Books.

Ladiges, W. 1983. *Coldwater Fish in the Home & Garden.* Tetra Press.

Randolph, E. 1990. *The Basic Book of Fish Keeping.* New York: Fawcett Crest.

Spotte, S. 1979. *Seawater Aquariums: The Captive Environment.* New York: Wiley-Interscience.

Vevers, G. (translator). 1973. *Dr. Sterba's Aquarium Handbook.* London: Pet Library, Ltd.

Anatomy of Tropical Fishes

Howard E. Evans

There are many shapes and sizes of tropical fishes, from the tiny mosquito fish (*Gambusia affinis*) of Texas, less than one inch (2.5 centimeters) long, to the large pirarucu (*Arapaima gigas*) of South America which reaches a length of 8 feet (2.4 meters). The smallest fish in the world is probably *Pandaka pygmaea* of the Philippines, which is mature at 0.4 inches (10 millimeters), and the largest is the whale shark, *Rhincodon typus,* measured at 45 feet (13.7 meters). For a comprehensive summary of the fishes of the world, see Nelson (1984).

The shape of the body usually is indicative of the habitat or behavior of the fish. Bottom-living fishes and fishes that cling to rocks in swift streams usually are flattened dorsoventrally, as are rays (*Dasyatis*) and many catfish (*Plecostomus*). Bottom-living fish that burrow or enter the substrate are usually serpentine or elongate, as is the kuhli loach (*Acanthophthalmus*). Mid-level fish are either "compressed" (flattened from side to side) as are tetras, or they are "fusiform" (torpedo-shaped) as are pencilfish and killifish. Extreme examples of compressed fish are the angelfish (*Pterophyllum*) and the sole (*Achirus*). The latter case is atypical, since all soles and flounders begin life as normally shaped fish but adapt to life on one side as they grow older. Although the eyes migrate to one side (which we call the dorsal surface instead of the right or left side), the fins, mouth, and oper-

cular flaps retain their original positions. There are thus both right-sided and left-sided flatfish, the side being determined by the position of the eyes. The flounders and the dabs have their eyes and color on the right side.

In addition to these common shapes, there are the truncate cowfish, globiform puffer, filiform pipefish, and bizarre seahorse (*Hippocampus*), not to mention the froglike mudskipper (*Periophthalmus*) with its bulging eyes, which spends so much time hopping around out of water.

Fins and Locomotion

The shape and construction of the body determine the locomotor ability of the fish to a great extent. Stiff, boxlike fishes encased in plates or covered by spines, such as the *Corydoras* catfish or the seahorse, rely entirely upon their fins for locomotion and are therefore rather slow. Not all fishes have multiple fins. In several, the fins on the median plane fuse so that the dorsal fin is continuous with the caudal and anal fins. Standard fishes have dorsal, caudal, and anal fins, and paired pectoral and pelvic fins. A small adipose fin lacking fin rays is present behind the dorsal fin on catfish, salmonids, and some tetras. Barbels, or whiskers, are named according to their site of origin.

Caudal fin: Rapid propulsion usually is a func-

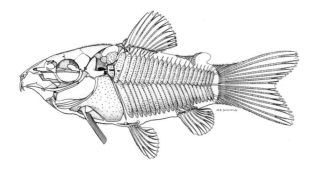

The skeleton of an armored catfish (*Corydoras*). Note the bony plates with spines covering the body and making it quite rigid.

tion of the caudal fin, or tail, aided by the pectorals in leaping. Serpentine or elongate fishes rely on an undulatory motion of the entire body for rapid movement. Fast-moving fishes usually have a forked tail and a narrow caudal peduncle with stabilizing finlets or lateral ridges, as are seen on mackerel. The tails of small fishes may be evenly or unevenly forked or may bear a ventral extension, as in the swordtail, or a median extension, as in the emperor tetra. There are round tails, square tails, triple tails, and veiltails, most of which can be modified by selective breeding.

Directional terms, names of fins and barbels, and planes of the body in a channel catfish (*Ictalurus punctatus*).

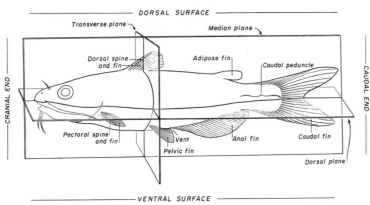

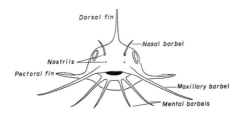

Pectoral fin: In addition to locomotor functions, the pectoral fins are used to counteract the forward movement of the body caused by the jet effect of water being forced out of the opercular opening. Enlarged pectoral fins are seen in fishes that leap out of the water, as do hatchetfish (*Gastropelecus*) and *Carnegiella.*

Pelvic fin: The pelvic fins stabilize the body against pitch and are used to counterbalance lift. They also produce roll when required. Pelvic fins may be located under the throat, at mid-body, or even further back. In gobies, the pelvic fins form a suction cup or a pedestal upon which they rest. The mudskipper "stands" upon the pelvic fin pedestal and uses the pectoral fins to hop when on land. The pelvic fins of the gourami and angelfish consist of a few elongated rays and function as tactile feelers.

Dorsal fin: The dorsal fin is usually prominent and serves as a stabilizer during forward motion. It can be used as a brake when the fish curls its caudal end. In many fishes the dorsal fin is erected for threat or courtship display and may be quite large or colorful, as in the Siamese fighting fish or sailfin mollies. In some marine fishes, such as the triggerfish (*Balistes*) and seahorse (*Hippocampus*), the dorsal fin is the primary locomotor fin. One or more spines may be associated with the dorsal fin. Some of these may be isolated from one another, as in the marine and freshwater sticklebacks (family Gasterosteidae), which may have three to sixteen isolated spines. These spines serve a defensive function and can inflict a painful wound. The spines of catfish are often accompanied by a poison gland.

Anal fin: The anal fin always is located behind the vent or anal opening, whichever is present. Like the dorsal fin, it serves as a stabilizer, and in some tetras it is very large.

In the live-bearing family Poeciliidae, which includes guppies, platys, and swordtails, the male has a rodlike anal fin called a gonopodium that serves as a copulatory organ. The presence of this spe-

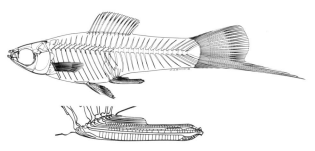

Skeleton of a male swordtail (*Xiphophorus helleri*). Inset shows structure of the anal fin, which is modified as a gonopodium for insemination.

cialized fin in guppies and platys enables one to recognize the male easily. (See also under "The Reproductive System.")

The Integument

Skin: The skin of fishes, as in all vertebrates, consists of an epidermis and a dermis. The epi-

Skin and scales of the swordtail (*Xiphophorus helleri*). A: Surface view with one scale removed; note that epidermis remains attached to the scale. B: Section to indicate a rupture of the scale pocket with the loss of a scale. Inset shows the underside of the scale with pigment-bearing melanophores in the dermis. C: Section to show a regenerating scale. Note that the epidermis has healed, closing the scale pocket. Growth and ossification of the new scale in the dermis has begun.

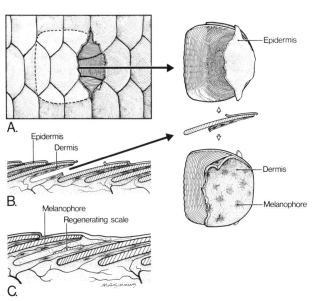

dermis is very thin (six to eight cell layers) and has unicellular mucous glands with a network of fine capillaries. The mucous glands of both the male and female discus (*Symphysodon*) produce a whitish secretion upon which the young fry feed for about three days before they start to take live food. The lungfish (*Protopterus*) uses skin mucus to enclose itself in a thin cocoon for aestivation in the mud during the dry season. Some parrotfishes (*Scarus*) enclose themselves in a cocoon of mucus every night while resting.

In addition to mucous glands, taste buds are widely distributed in the epidermis of many fishes. These taste buds transmit impulses over branches of the facial nerve to the hindbrain. There also may be pigment cells and alarm cells that produce pheromones. In silvery fishes, there are few if any pigment cells in the epidermis, whereas in others such as the red oscar (*Astronotus*), the pigment is primarily in the epidermis. The epidermis is easily broken when a fish is handled or is bitten by another fish, rendering the dermis vulnerable to infection.

The dermis usually contains many melanophores and a rich plexus of blood vessels and nerves. Scales develop in the dermis and lie within pockets close to the surface. When a scale is lost, that portion of the epidermis attached to its free surface is lost with it. As the epidermis grows and the wound heals, osteogenic cells remaining in the ventral wall of the scale pocket initiate the formation of a new scale.

Scales: The scale of a teleost fish is usually thin and flexible. Since it is a dermal scale, rather than an epidermal scale as in reptiles, it is not shed regularly, although it can be replaced if lost. It is composed of a type of bone tissue and is constructed in a manner that resembles plywood. A typical scale has two primary layers: a bony or hyalodentine surface layer that may bear spines, ridges, or grooves; and a deeper basal plate of woven lamellar composition.

The bony layer is usually acellular, although typical osteocytes may be seen in several families of fishes (Mormyridae, Gymnarchidae, Osteoglossidae). The lamellar layer of the basal plate consists of directionally alternating strata (as many as twenty-five layers) of parallel collagen fibers mineralized to vary-

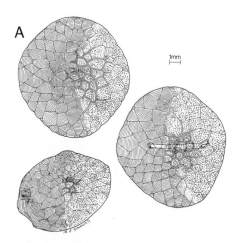

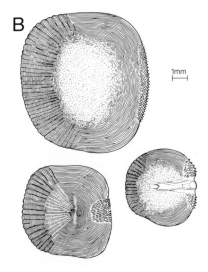

A: Scales of an arowana (*Osteoglossum bicirrhosum*), one of five living species of ancient bony-tongued fishes. These large scales lack radii and circuli. B: Scales of an oscar (*Astronotus ocellatus*), showing radii, circuli, and cteni. The exposed surfaces of all scales are on the right.

ing degrees and incorporating occasional or numerous calcified concentrations called Mandl's bodies. Although the basal plate of the scale is formed primarily by horizontal layers, several fishes (barbs, minnows, and carps) show transverse fibrils which divide the horizontal bundles into more or less regular segments. Such transverse fibrils may serve to spread mineralization through the basal plate and to bind the stratified layers together.

Most scales have concentric ridges, or circuli, in the bony layer that are deposited throughout life around the margins of the scale. When several circuli are close together, a ring, or annulus, is formed. These annuli vary in width or density due to metabolic growth stimulation or inhibition, and thus can be used to determine the age of a fish if there is some environmental periodicity in the annual cycle. The annuli are crossed by grooves, or radii, on most scales.

The radii, which appear as spokes from the center of the scale, are usually more numerous on the cranial or covered portion of the scale. There is a peculiar lack of mineralization in the lamellar layer beneath the radii of the scale, giving the appearance of clefts. Some species have many more radii on their scales than do others, and some lack radii entirely.

The cranial portion of each scale is overlapped by scales ahead and to the sides of it. On their exposed surface, scales may be smooth (cycloid) or spiny (ctenoid). Usually only the caudal margin of the scale bears spines or appears comblike. In

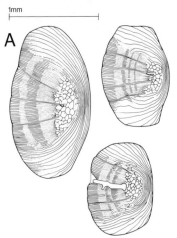

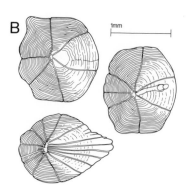

A: Scales of the black tetra (*Gymnocorymbus ternetzi*), have a cellular center and lack complete circuli. B: Scales of a Sumatra barb (*Barbus tetrazona*), have complete circuli and prominent radii in all quadrants.

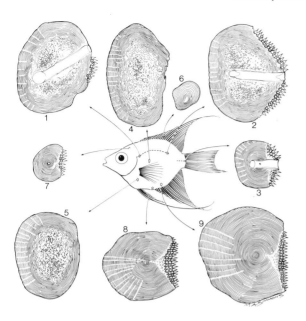

Types of scales seen on an angelfish (*Pterophyllum*). All of the scales show growth rings or circuli. 1, 2, 3: Ctenoid lateral line scales enclosing a canal. 4, 5: Cycloid scales with a large central focus. 6: Cycloid scale with a small focus and no radii. 7: Ctenoid scale with a small focus and no radii. 8, 9: Ctenoid scales with a small focus and long radii.

angelfish, the spines seen on the caudal edge of the scale form in such a manner that the terminal spine of each radial row is the youngest. As each new spine forms, its base (which lies beneath the

Scales of a swordtail (*Xiphophorus helleri*), at the middle of its body, show pattern of overlap. In the central shaded area, five scales overlap.

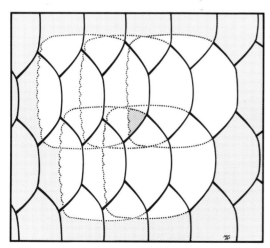

tip of the older spine) is indented and cradles the older spine. When observed by transmitted light, the apex of each spine appears to be capped by the successive spine, but that is only an illusion.

A single fish may bear several kinds of scales. The size and shape of the scales may vary on different parts of the body, and sometimes there are scaleless areas, particularly on the head. Scales on particular parts of the head and body may have sensory canals passing through them. Each species of fish has its typical pattern of sensory canals, which begin as surface structures in the embryo and are gradually enclosed by scales or bone. Most commonly there is a lateral line canal on each side of the body that begins on the head and extends to the caudal peduncle. Many fishes have only a portion of this canal present, and in some it is discontinuous or absent. The lateral line canal opens to the surface by one or more pores through the scale. This allows water to enter the canal and transmit pressure changes or vibrations via neuromasts that lie within the canal.

Not all fishes have scales. Most catfish (family Siluridae) and many species of eels lack scales. Genetic variants of the carp (*Cyprinus carpio*) that are incompletely scaled are called "mirror carp," and those without scales are referred to as "leather carp." Various percentages of spawn from a carp hatch will exhibit such abnormal scalation.

Scale replacement: When a scale is lost, the scale pocket fills with gelatinous material and the epidermis heals rapidly. In the center of the floor of the scale pocket, one or more scale papillae develop as an aggregation surrounded by uncalcified bone matrix called osteoid. Osteoblastic activity is seen on both surfaces of the developing scale as it enlarges to fill the scale pocket.

Regenerated scales can sometimes be recognized by a difference in size, conformation, or spacing between the annuli or radii. There may be multiple growth centers in the new scale or uneven ossification. In some minnows, such as the common shiner (*Notropis cornutus*), regenerated scales are easily recognizable because they lack the silvery appearance of an original scale. When melanocytes are numerous beneath a layer of reflecting guanin crystals, a silvery appearance results. Since the melanophores are most numerous on the underside of the scale and are lost with it, there is often a change in the color pattern of

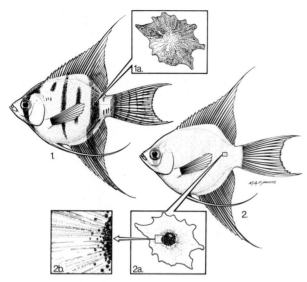

Angelfish (*Pteryophyllum*) have dark bars (1) when pigment is diffused in the melanocyte (1a), which is the usual condition when the fish is calm. When the fish is excited or in bright light, the color fades (2) because the pigment granules concentrate in the center of the melanocyte (2a). Pigment granules appear to move along microtubules (2b) under the influence of sympathetic nervous innervation. Note that the shape and size of the melanocyte remains the same regardless of the coloration.

the regenerated scale due to a difference in the number or distribution of new pigment cells that migrate into the area.

Color and pattern: The colors of fishes are produced by pigment cells in various combinations with the properties of refraction and reflection of light as it passes through these cells and scale structures. The various kinds of pigment cells, or chromatophores, differ in their chemical composition and appearance, although they have a common origin as endoplasmic reticulum. The most common chromatophores are the melanophores, with black or brown melanins contained in organelles called melanosomes. Xanthophores (yellow pigment cells) utilize pteridines synthesized in the cell and deposited as pterinosomes, whereas erythrophores (red pigment cells) concentrate carotenoids of dietary origin in vesicles of various sizes. Several types of pigment organelles may occur in a single cell.

Iridophores (guanophores) are not pigment cells. They utilize crystalline deposits of purines to produce structural colors by reflection, scattering, diffraction, or interference of light. They have also been called refractosomes.

Changes in color or pattern can be gradual or rapid. A developmental or morphological change in color is a slow process that depends upon the accumulation of pigment from food or water, an increase in the number of pigment cells (melanophores), or a change in the scale pattern. Physiological color change mediated by neural mechanisms can be slow or almost instantaneous. It results from the intracellular redistribution of pigment-containing organelles, or chromatosomes, within a pigment cell. When chromatosomes contract (aggregate) to the center of the cell, color fades. Normally this is a response to fright or to bright light. When pigment granules are dispersed in the pigment cell, the entire cell darkens and the overall effect is that the color pattern appears more intense. Pigment granules are transported along radial microtubules within the chromatophore. The cell maintains its shape

A female platy-swordtail hybrid, showing a band of concentrated melanocytes in the dermis beneath the scales. Where pigment cells are scarce or lacking, the scale outline is not visible. Excessive growth of macromelanophores on various sites of the head, fins, or body can result in melanomas or erythroblastomas (pigment-cell tumors) and the death of the fish. There is a genetic basis for this development.

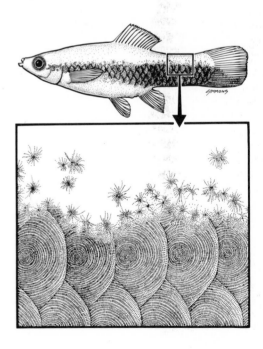

regardless of the aggregation or dispersion of the pigment granules.

Pigment-cell tumors: In 1932, Myron Gordon of the New York Aquarium was one of the first investigators to utilize hybrids of the Mexican swordtail (*Xiphophorus helleri*) and the platy (*X. maculatus*) to study melanosis, a neoplastic condition characterized by large spreading tumors in the skin. He showed (Gordon 1927) that the tumors had a hereditary basis and could be produced by crossing a normal female swordtail with a stippled and spotted male platy to produce a mottled platy-swordtail hybrid which frequently developed melanosis. Several varieties of hybrids in the aquarium trade have been named for their red or black benign or malignant color patterns.

There is the likelihood that one or more genes (oncogenes) are the causative agents of tumor formation in fishes. Several benign and several very malignant melanomas have been produced. The nature of these tumors suggests that they arise from cells of neural-crest origin. (For a review of normal and neoplastic melanophores in the skin of the swordtail, see Vielkind and Vielkind 1982.)

The Skeletal System

Structure of the head: The head consists of the skull, opercular bones, gill arches, and jaws, with the associated teeth, soft parts, and sensory organs such as the eye, ear, and brain.

The operculum is a muscle-controlled flap on each side of the head that covers the gill chamber and acts as a valve to open and close the pharynx for water and food intake or water and debris outflow. The flap is stiffened by several bones, the largest being the opercular bone, which articulates with others whose outlines are sometimes visible on the skin surface. Muscles associated with the opercular flaps can close the gill chamber (adductors) or open it (abductors). Abduction will also flare the "cheeks," as in threat displays or when allowing fry to enter for protection, as do some cichlids.

The branchiostegal rays are a series of flat bony bars that stiffen the branchiostegal membrane, which stretches from the lower margin of the opercular flap to the isthmus or midline of the throat. This structure functions as a gasket for better closure when the operculum is adducted and the pharynx is expanded for water intake. The right and left branchiostegal membranes may be fused along most of their length on the midline, thus restricting the exit from the gill chamber. The eel, seahorse, and *Plecostomus* provide extreme examples of this narrowing of the opercular exit. In fishes that must attach to the substrate in order to hold their position, water enters and leaves the gill chamber through openings on the opercular margin of the isthmus rather than through the mouth.

The skull: The skull of a fish has more bones (approximately 185) than are found in the skull of any other vertebrate. To understand the skull, it is best to consider the component parts as units.

The skull includes an inner, brain-containing neurocranium, and the branchiocranium, which consists of outer groups of articulating bones such as the jaws, the hyoid apparatus, and gill-bearing arches. Other less easily categorized bones include groups such as the two bones in the sclera of each eyeball and the circumorbital series around the eye that enclose a sensory canal. Not all fishes have the same number of individual bones; some bones fuse with one another, and others are lost in development or evolution.

1. The neurocranium, or braincase, of a fish, when viewed by itself, does not sug-

Bones of the jaw and opercular flap. The maxilla in most fishes is toothless and does not form a margin of the mouth opening. The premaxilla and mandible are usually toothed, except in cyprinid fish. The branchiostegal membrane and its rays serve as a gasket for closing the gill opening.

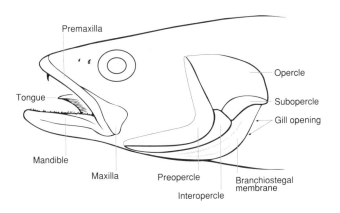

Premaxilla
Opercle
Subopercle
Gill opening
Tongue
Mandible
Maxilla
Preopercle
Interopercle
Branchiostegal membrane

gest the appearance of a fish's head. It is rather triangular in lateral view. The base or horizontal axis is formed by the parasphenoid and basioccipital bones. The dorsum is formed by the ethmoid bone rostrally, followed by the frontal, parietal, and supraoccipital bones. If a fish has a high crest on the skull, this is formed by the supraoccipital bone. The caudal portion of the neurocranium is formed largely by the exoccipital, basioccipital, and otic bones that enclose the inner ear.

The otic capsule contains the membranous labyrinth of the inner ear. This labyrinth consists of three semicircular ducts: a utriculus, a sacculus, and a lagena. Each of the chambers has an ear stone, or otolith, which functions for sound reception (see also "Balance and hearing").

The basicranial region is formed primarily by the basioccipital bone, which articulates with the vertebral column. Rather than a freely articulating joint, there is an opisthocoelous facet on the basioccipital bone that resembles one end of a vertebral centrum. In minnows and carps (the family Cyprinidae), the basioccipital bone has a ventral median plate that bears a cartilaginous pad against which the "pharyngeal teeth" (see later in this section) bite. When this pad dries, it becomes very hard and is spoken of as a "carp stone," "Karpfenstein," or *"la meule."* The basioccipital bone has a caudal extension that forms a ventral arch through which the dorsal aorta passes.

2. The branchiocranium consists of the jaws, the hyoid apparatus, and the gill basket.

The upper jaw, palatine arch, and lower jaw on each side of the branchiocranium comprise the oromandibular region. In the cyprinid fishes (barbs, carps, and minnows), the upper jaw consists of a toothless premaxilla and maxilla. The premaxilla in most fishes excludes the maxilla from the mouth opening. The upper jaw of many fishes is protrusible, as in the goldfish.

The hyoid apparatus consists of the hyomandibular bone, symplectic bone, the opercular series, the tongue elements, and the branchiostegal rays.

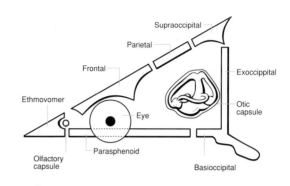

Schematic view of the fish neurocranium or braincase. This fusion of skull bones houses the brain and membranous labyrinth of the ear. The remaining bones of the head cover or articulate with the neurocranium.

Sense organs and brain of a spotted sea-trout (*Cynoscion nebulosus*) in dorsal view. The dashed outline below the semicircular ducts indicates the position of the sacculus and lagena of the inner ear.

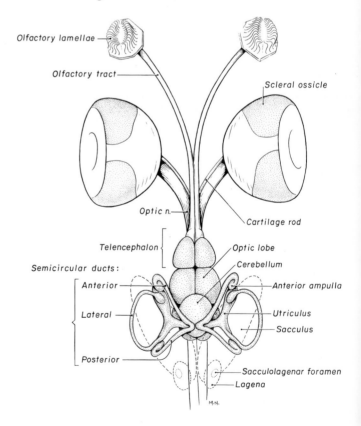

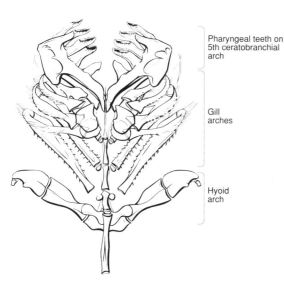

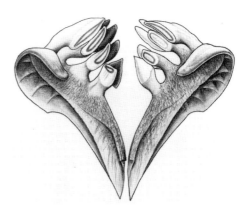

The pharyngeal skeleton of a golden shiner (*Notemigonus chrysoleucas*) in dorsal view. A median element of the hyoid arch supports the tongue. Four gill arches bear rakers and filaments. The modified fifth gill arch bears the pharyngeal teeth.

The hyomandibular bone is an elongated bone which functions as the suspensory element between the skull and the jaws, hyoid apparatus, and opercular bones. The upper end of the hyomandibular bone articulates in a socket on the skull. The lower portion of the hyomandibular bone is tightly bound to the preopercular bone, and its ventral end gives rise to a stout tendon which passes to the symplectic bone. It is upon

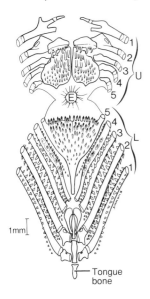

Upper and lower pharyngeal teeth of an angelfish (*Pterophyllum*) as seen through the open mouth. E: esophagus; L 1-4: gill arches; L 5: lower pharyngeal teeth on fused right and left ceratobranchials; U 1-5: upper ends of gill arches. Note the paired, tooth-bearing upper pharyngeal plates.

Pharyngeal teeth on the fifth ceratobranchials of a goldfish (*Carassius auratus*) in dorsal view with the rostral end at the bottom. There are four functional teeth on each side, and the tooth replacement sequence appears to alternate left and right sides. The median (first) tooth is always slightly ahead of the other three. In this illustration, the replacement teeth, on the right side of the fish, are young and stain darkly. On the left side, the replacement teeth are older and covered by nonstaining enamel. The pointed first tooth on the left arch has already moved into functional position, and the remaining three functional teeth will soon be shed and replaced by those behind them. This cycle continues throughout the fish's life.

the medial side of this tendon that the interhyal bone, which suspends the tongue apparatus, is attached. The symplectic is a splintlike bone which is wedged into the posterior notch of the quadrate bone. On the posterior margin of the hyomandibular there is a hemispherical opercular condyle that fits into a socket on the opercular bone. This joint permits movement of the opercular flaps. Other bones within the opercular flap are the subopercular and interopercular bones.

The hyal elements, or tongue apparatus, support the floor of the mouth and pharynx. A small nodular interhyal bone represents an ossification in the tendon connecting the hyoid arch to the cartilaginous union between the hyomandibular and symplectic bones. The next bone medially is the epihyal, which in turn attaches to the angular. The ceratohyal is the longest bone of the series, followed by the dorsal and ventral hypohyals, and, on the midline, by the basihyal within the tongue and urohyal behind it.

The branchial region consists of the gill-bearing arches and their associated gill rakers and pharyngeal teeth. A typical arch is composed of

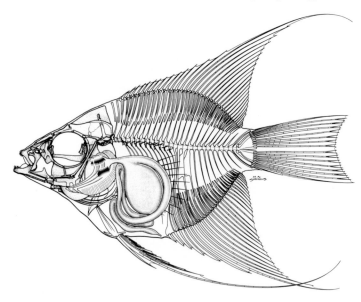

The skeleton and digestive tract of an angelfish (*Pterophyllum*). Left gill arches have been removed to show the position of pharyngeal teeth at the entrance to the esophagus. The stomach is saclike, and there are three loops of intestine.

a series of bones on each side: the basibranchial, hypobranchial, ceratobranchial, epibranchial, and pharyngobranchial bones. Frequently the fifth arch consists of a large tooth-bearing ceratobranchial bone (as in minnows and carps) or a combination of tooth-bearing ceratobranchials and pharyngobranchials that oppose each other (as in cichlids). These latter structures are spoken of as the "pharyngeal teeth" or the "pharyngeal mill."

Sound-transmitting bones (Weberian ossicles) of the horned dace (*Semotilus atromaculatus*). Several families of fishes have one or more movable elements that pivot on the vertebrae and serve to conduct sound from the gas bladder to the inner ear via pressure on a perilymphatic sac.

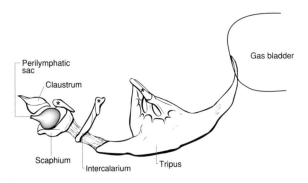

The gill rakers are arranged in two rows on the medial side of the gill arch, and the gill filaments are attached to the lateral side of the arch. The entire gill basket is tightly bound to the roof of the pharynx dorsally and to the hyoid apparatus ventrally.

The vertebral column: The length of the vertebral column and the number of vertebrae are not always related. Length is dependent upon the shape of the fish, but the number of vertebrae may be affected by the temperature of the water in which the fish develop. Eels have the most vertebrae, whereas discus, angelfish, and other short fishes have the fewest. When comparing two fishes of the same species from different habitats, the fish which developed in colder water will be the most likely to have more vertebrae.

Fish vertebrae are amphicoelous, having a spindle-shaped centrum with a central canal through which the notochord passes. Between vertebrae, the notochord expands to fill the space. Vertebrae are attached to each other by a fibrous ring at the margin of the centrum which allows only limited movement. Above the centrum are the neural arch and spine, which enclose the spinal cord. The dorsal fin has anchoring interneural bones intercalated between the neural spines. Below the centrum are ribs on each side in the abdominal region and hemal arches in the caudal region. The hemal arch encloses the caudal artery, which is a continuation of the dorsal aorta and the caudal vein, which returns blood to the heart. A ventral continuation of the hemal arch is the hemal spine. Anchoring the anal fin in the body musculature are interhemal bones intercalated between the hemal spines.

Vertebrae at the cranial end of the vertebral column may be modified to cradle the gas bladder or to serve as sound-transmitting ossicles. All carps, characins, catfish, and loaches (ostariophysid fishes) have some type of linkage between the gas bladder and the inner ear, mollies (Schröder 1964). Thus there is variable expression of lyretail.

The veiltail molly has a large caudal fin and all of its other fins are also enlarged. A single female veiltail was discovered in a fish pool (Wolfsheimer 1965; Wood 1968). Veiltail is due to a single dominant gene (Norton 1974) with variable expressivity, the tail varying in size and shape. Many veiltail molly males are unable to breed because of their deformed gonopodia. Crossing a veiltail female with a lyretail male produces four kinds of offspring: veiltail, lyretail, lyreveil (having both veiltail and lyretail), and common (Norton 1974). Variable expression occurs not only in veiltails but also in lyreveils, which have genes for veiltail and lyretail. A lyreveil with a large tail has an enlarged tail with short extensions at the upper and lower edges of the tail. A lyreveil with a small tail looks very much like a lyretail, having a somewhat larger tail than in a lyretail. Genetic testing by crossing with a common is necessary to accurately differentiate between a lyretail and a small-tail lyreveil. The genes for lyretail and veiltail are nonallelic (Norton 1974).

Variable expressivity of pigment-pattern genes in *Xiphophorus* is responsible for numerous variations in domesticated platys and swordtails. For example, color intensity of red platys and swordtails is influenced by modifiers of the platy gene for red body. The red dorsal fin color of the platy (*X. maculatus*) spreads over most of the body due to modifiers of red dorsal that are added by crossing a red-dorsal platy with a wild swordtail (*X. helleri*).

The expression of two of the tail-spot alleles, comet and twin-spot, in the platy (*Xiphophorus maculatus*) is changed by a different modifier of each (Gordon 1956). A dominant modifier (*Cg*) changes twin-spot (two black spots at the base of the tail) to Guatemala crescent, a broad crescent at the base of the tail. Guatemala crescent has been found in platy populations only of the Lake Peten region in Guatemala. However, the same modifier also occurs in the swordtail (*X. helleri*) (Gordon 1956; Atz 1962). Guatemala crescent did not appear in hybrids of twin-spot *X. maculatus* crossed with *X. milleri, X. couchianus,* and *X. nigrensis* (Kallman 1975). The *Cg* modifier has no effect on the other tail-spot patterns. Also, in *Xiphophorus,* a dominant factor (*E*) changes the comet pattern to wagtail, but has no effect on any other tail-spot patterns tested. In comet, the upper and lower edges of the tail are

black, called "twin bar" in the trade. All of the fins are black in a wagtail. The *E* gene occurs in *X. helleri,* not in *X. maculatus* (Gordon 1946), and natural populations of *X. helleri* are homozygous for *E.* Another modifier affects a tail-spot pattern of *X. variatus* (Borowsky 1984).

Yellow and red pigment patterns of *Xiphophorus* also may vary in their expression when they are incorporated into the genetic makeup of other species of *Xiphophorus.* Yellow patterns of *X. pygmaeus* become red patterns when introduced into *X. maculatus* (Zander 1969), but the red dorsal (*Dr*) gene of a Jamapa strain of *X. maculatus* produces yellow dorsal fins in *X. montezumae* (Zander 1969), now known as *X. nezahualcoyotl* (Rauchenberger, Kallman and Morizot 1990).

Lethal or deleterious genes: In guppies, the recessive gene for albinism is deleterious when homozygous (Haskins and Haskins 1948). Insemination of a female by an albino male usually results in only one small brood instead of the multiple broods common in poeciliids. Albino guppies are more sensitive to poor environmental conditions than are non-albino guppies, and their life-span is shorter than that of wild-type guppies. The F2 from an albino crossed with a wild-type would theoretically be in the ratio of 3 wild-type :1 albino. The actual ratio obtained by Haskins and Haskins (1948) was 51.1 wild-type :1 albino. However, females varied in the percentage of albino offspring they produced.

Albinism also is harmful in swordtails. Albino and wild-type embryos were present in the expected ratio in dissected females that previously had produced broods having fewer than the expected percentage of albino fry (Gordon 1942). Albino embryo mortality occurs shortly before or after birth. Some swordtail females produce close to the expected percentage of albino offspring, whereas other females produce fewer than the expected percentage of albinos.

In guppies, recessive lethals linked to color-pattern genes on the Y chromosome (a sex chromosome) prevent the production of offspring homozygous for these genes (Winge and Ditlevsen 1938; Haskins et al. 1970).

Inheritance and Sex

Sex-limited factors: Factors that can occur in both sexes, but are expressed in only one sex, are called sex-limited.

The thalamus proper has a very small dorsal region and a larger ventral region. Optic projections from the retina go to the ventral thalamus as well as to the optic lobe of the mesencephalon. The hypothalamus is the largest portion of the diencephalon in fishes and can be seen readily on the ventral surface of the brain. The hypothalamus receives gustatory and olfactory impulses as well as optic connections from the tectum.

The mesencephalon is proportionately large in fishes and forms prominent optic lobes. The superficial layer of the optic tectum receives most of the projecting fibers from the retina, and in a fish lacking eyes, the optic lobe would be noticeably reduced. However, deeper portions of the optic tectum, even in blind fishes, are still represented because of

The brain of a goldfish (*Carassius auratus*). The large vagal lobes of the medulla indicate the presence of many taste buds in the mouth. Goldfish reject food that does not taste good to them.

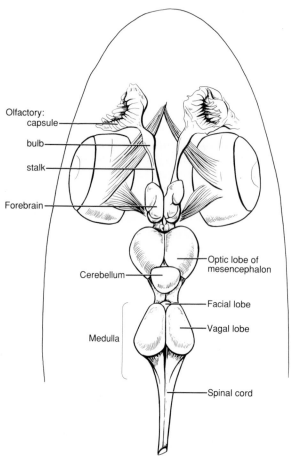

Olfactory:
  capsule
  bulb
  stalk

Forebrain

Cerebellum

Optic lobe of
mesencephalon

Facial lobe

Vagal lobe

Medulla

Spinal cord

somatic sensory projections from the brain stem and visceral projections from the hypothalamus. Thus, the optic tectum is essentially a correlation center that discharges to lower centers and to the cerebellum through relays in the tegmentum or reticular formation.

The cerebellum (metencephalon) is large in most fishes and particularly so in electric fishes. It often has an enlarged medial portion, called the valvula cerebelli, projecting beneath the optic tectum. Impulses from lateral line nerves in cranial nerves V, VII, IX, and X pass to the cerebellum. Auricular lobes, associated with the vestibular apparatus for balance, may be visible externally in some fishes.

The medulla or hindbrain (myelencephalon) indicates most noticeably which gustatory sensory modalities are utilized by a fish when feeding. On the dorsal surface, the central nuclei of cranial nerves carrying afferent impulses from taste buds are enlarged and visible grossly. The visceral afferent column at the level of the facial nerve, which receives impulses from taste buds on the skin and barbels as well as communis fibers from the lower body surface, is seen as a median facial lobe in minnows or as paired median facial lobes in catfish. The enlargement of the medulla at the level of the afferent glossopharyngeal and vagal nerves forms the vagal lobes of the brain and represents the input of taste buds on the palatal organ and pharynx. Thus a fish that recognizes its food on the skin or barbels will have a large facial lobe, whereas one that samples its food in the mouth will have large vagal lobes. Some fishes (such as shiners) do neither and have very small facial and vagal lobes, whereas others do both and show enlargements of both.

Cranial nerves: There are ten cranial nerves in fishes, designated by Roman numerals. Olfactory nerves (I) are very short and pass from nasal sac mucosa to the olfactory bulb. Optic nerves (II) are large and stiffened by a cartilage rod; at the level of the telencephalon they cross completely. Oculomotor nerves (III) innervate the dorsal, ventral, and medial rectus and the ventral oblique muscles. Trochlear nerves (IV) supply the dorsal oblique muscles. Trigeminal nerves (V) innervate the jaw muscles and are sensory to the pelvic fin "feelers."

Abducent nerves (VI) supply the lateral rectus muscles of the eye. Facial nerves (VII) are sensory to taste buds on the skin of the head, body, and fins. Vestibulocochlear nerves (VIII) receive impulses from the ampullae of the semicircular ducts, utriculus, sacculus, and lagena. They are joined by fibers from the acousticolateralis system. Glossopharyngeal nerves (IX) are sensory to the pharynx. Vagal nerves (X) innervate taste buds of the palatal organ, pharynx, and the musculature of the gut.

The spinal cord: The spinal cord of fishes usually extends through the entire length of the vertebral canal. In a fish with over 200 vertebrae, such as an eel, the spinal cord is of considerable length and the spinal nerves exit segmentally along its length. On the other hand, in short fishes with fewer than twenty-five vertebrae, such as the puffer, the spinal cord is extremely short and only a cauda equina of spinal nerves continues down the vertebral canal. The ocean sunfish (*Mola mola*) is an extreme case where the length of the brain almost equals the length of the cord.

Dorsal and ventral nerve roots join outside of the vertebral canal, and the dorsal ganglion also lies at this point. The dorsal ganglion is often very large in fishes and is supplied by a capillary plexus. The ganglion has bipolar and unipolar neurons as well as intermediate types. The dorsal and ventral roots do not always leave the cord at exactly the same level. Visceral efferent fibers course in both dorsal and ventral roots.

Some fishes have a few pairs of nerves between the branchial nerves and the first spinal nerves. These are referred to as occipital or occipitospinal nerves, and they emerge from the skull or from behind the skull. In most instances these occipital nerves join the first two to four spinal nerves and form plexuses which supply hyoid and pectoral girdle musculature.

At the caudal end of the spinal cord in fishes, there is a caudal neurosecretory system consisting of vessels (urophysis) and glandular cells interrelated in some manner and of undetermined function.

Sensory Activitites

The nervous system of fishes operates on a reflex level, wherein the inputs of olfaction, taste, vision, touch, balance, and hearing interact without much cerebration. Since a fish's major activity is feeding, and its major concern is being eaten, its sensory system is perfected to maximize the former and minimize the latter. Sense organs are numerous and range from very simple free nerve endings for chemical reception to very specialized electroreceptors capable of low-voltage recognition.

Touch: General tactile (trigeminal) sensation is well developed in fishes and is often associated with modified pelvic fin rays such as those seen in gourami, angelfish, and sea robins. These fishes utilize their filamentous fins as "feelers" along the bottom or as "early warning" devices when extended from a hiding place. The barbels of a catfish serve for both touch and taste. The sensation of touch is carried by the trigeminal or fifth cranial nerve, and taste by the facial or seventh cranial nerve on the barbel. There are

The brain and cranial nerves of a spotted sea-trout (*Cynoscion*), a drum that feeds by sight.

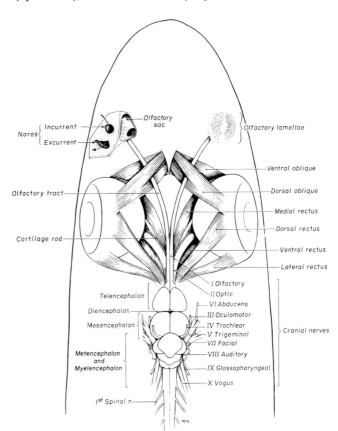

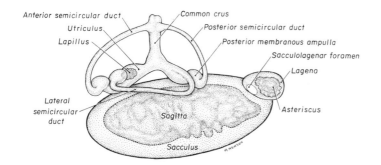

Membranous labyrinth of the inner ear of a spotted sea-trout, showing the three ear stones or otoliths in their respective chambers.

also numerous pressure receptors in the lips and pharynx.

Taste: Gustatory sensations are received by taste buds which look very similar to those of other vertebrates (Kapoor, Evans, and Pevzner 1975). However, they may be located anywhere on the fish body or fins as well as within the oral cavity. Taste buds are particularly numerous on the barbels; there may be in excess of 1,600 buds per square centimeter (over 10,000 buds per square inch) in the brown bullhead (*Ictalurus nebulosus*). When the barbels of a catfish are extended to each side, they act as triangulating devices to home in on a source of food molecules that are being disseminated in the water. If the food source is directly ahead of the fish, the sensations being received are equally strong from both barbels, and the fish swims ahead. If the stimulus is stronger in one barbel than the other, it indicates to the fish that a turn is being made away from the stimulus, and a compensatory movement is made to head for the target. If a barbel is lost, it will regenerate. While this is taking place, the fish circles toward the side of the lost barbel since it does not perceive any reduced stimulus from that side.

There is a difference in the innervation of taste buds on the surface of the skin versus those within the oral cavity that is reflected by the size of the receptor nuclei in the brain. Taste buds of the skin and barbels are innervated by the facial nerve (VII) and transmit to the facial lobe of the medulla, whereas those within the oral cavity and pharynx are innervated by the vagal nerve (X) and transmit to the vagal lobes on each side of the medulla (see also under "Cranial nerves"). A fish

that characteristically mouths its food and rejects items it has engulfed, such as a goldfish, will be found to be utilizing vagally innervated taste buds. Such a fish will have greatly enlarged vagal lobes of the brain. Fishes that utilize taste buds on the barbels and skin, such as catfishes, have enlarged facial lobes.

Smell: Olfaction serves a social function in fishes for recognition of status within the group or it serves to warn of the presence of a predator requiring fight or flight reactions. The olfactory sac has an incurrent and excurrent opening separated by a baffle that projects above the surface and directs the water current. In teleost fishes there is no connection between the olfactory sac and the pharynx. Within the olfactory sac there is an olfactory rosette consisting of many lamellae covered by receptor mucosa. Below the mucosa there is an olfactory bulb of the brain which receives the short primary olfactory neurons of cranial nerve I.

Sight: The visual system is very important for most fishes, but some are able to get along quite well without eyes, as is shown by the blind cave characin (*Astyanax*), which lives and breeds well in captivity as well as in the wild. Fishes active at night are characterized by large eyes, a common feature of nocturnal reef fishes. Fishes of caves, muddy rivers, or abyssal depths often have very reduced eyes.

All fishes are nearsighted, as can be surmised from the spherical lens and the weak accommodation mechanism of the retractor lentis muscle. The sclera of all fishes has a pair of bony plates or ossicles to stiffen it. There are a few modifications of the eye in fishes whose behavior requires vision both in air and water, such as the four-eyed fish (*Anableps*), which has a constriction between the dorsal and ventral portions of the cornea, and the mudskipper (*Periophthalmus*), with its bulging, froglike eyes.

Balance and hearing: It is difficult to separate the sense of balance from that of hearing, because several parts of the ear serve both functions. The inner ear within the otic bones of the skull is relatively large. It consists of three semicircular ducts with ampulae: a utriculus, a sacculus, and a lagena. Most often, all of these

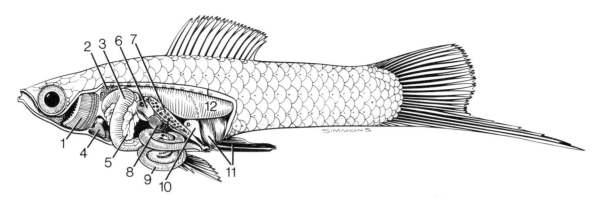

Viscera of a male swordtail (*Xiphophorus helleri*). 1: gill lamellae; 2: head kidney; 3: stomach; 4: heart; 5: liver; 6: spleen; 7: testis; 8: gallbladder; 9: intestine; 10: urinary bladder; 11: gonopodium and its muscles; 12: gas bladder.

chambers, which are filled with endolymphatic fluid, communicate with each other. Within the utriculus, sacculus, and lagena, there are sensory cells with sterocilia upon which rests an otolith. The otoliths are free to move and respond to vibration or to changes in position and acceleration. Any movement of the otolith is transduced as an auditory impulse via the eighth cranial nerve. The sacculus is the largest chamber in drum fish and the saccular otolith, called the sagitta, is likewise large. It is the ear stone most frequently used for determining the age of a fish and can be seen clearly on a radiograph or within a cleared skull. As would be expected, those fishes that produce sounds as part of their courtship behavior, such as drums, have the largest otoliths.

The lateral line sensory canals with their enclosed neuromasts also transmit the reception of vibrations and water movement to the acoustico-lateral system. Present on the head and along the body, these water-filled canals have pores that open to the surface through the skin or scales. They are particularly well developed in cave fishes.

Other important sensory organs include the electroreceptors of mormyrids, gymnotids, and sharks, which appear as jelly-filled pit organs, rosettes, or ampullae of Lorenzinni.

The Digestive System

The mouth: The mouth is the limiting factor determining the size and type of food eaten.

Toothless fishes such as cyprinids (barbs, danios, goldfish) gulp their food and lacerate it in the pharynx by means of pharyngeal teeth before swallowing. Fishes with teeth on their jaws, such as cyprinodontids, cichlids, and characins, use their teeth to nibble and tear bits of food from larger pieces. They may also have upper and/or lower pharyngeal teeth in the pharynx for further maceration. Some fishes have a very wide mouth opening, as do catfish and cichlids, whereas most tetras and cyprinids have small mouths. In general, bottom feeders have protrusible jaws on the underside of the head, midlevel feeders have terminal nonprotrusible jaws, and surface feeders have transversely oriented jaws that may or may not be protrusible. One of the most restrictive mouth openings is seen in the seahorse, which has small toothless jaws fused into a tube that is capable of suction but little else.

The lips: The lips may be thick, thin, or expanded into a suction pad, as in *Plecostomus*. In some fishes the lips are so thick that they hide the teeth (cichlids); in others, such as the piranha or bucktooth tetra (*Exodon*), the teeth are clearly visible externally even when the mouth is closed. The lips usually have several ridges or papillae on their inner surface and may have papillae along their free margins. When the papillae are numerous, the lips appear thick. There is a distinct labial groove between the lip and the jaw. On the labial margin of the jaws the teeth are arranged in rows that are partially or completely hidden by papillae.

The barbels: Extending from the lips may be a pair or more of barbels. Barbs, carps, and danios

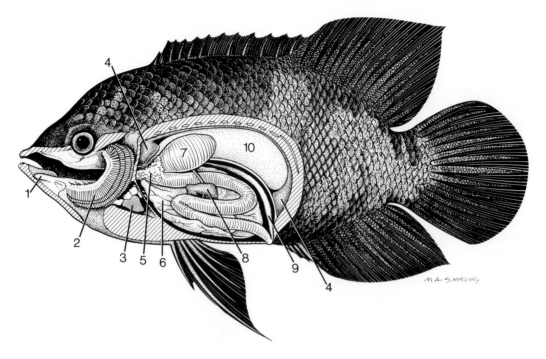

Viscera of a male oscar (*Astronotus*). 1: tongue; 2: gill lamellae; 3: ventricle of heart; 4: kidney; 5: pancreas; 6: liver; 7: stomach; 8: spleen; 9: testis; 10: gas bladder.

have small barbels without an internal stiffening rod. Goldfish (*Carassius*) have no barbels, whereas carp and koi (*Cyprinus*) have two pairs of small barbels. It is interesting that hybrids of carp and goldfish usually have only one pair. Barbels are gustatory and tactile sensory structures by virtue of the taste buds and the free nerve endings that they bear. In catfishes the long barbels are moved by muscles and they are used as long-distance directional sensors, in contrast to the trailing or protruding barbels which serve as close-range sensors for the majority of fishes.

The oral cavity and valves: The upper and lower oral valves separate the papillate labial cavity from the smooth oral cavity. Oral valves are thin flaps which extend from one angle of the jaw to the other. They are in apposition when the jaws are closed and thus serve to occlude the oral cavity when water is being forced out of the opercular openings.

The oral cavity is the region between the oral valves and the gill chamber. Its lining is smooth, and on the ventral surface lies the tongue, which may be fleshy or covered with teeth. There is often a frenulum extending from the tongue to the level of the oral valve.

The pharynx: The pharynx is the chamber be-

tween the oral cavity and the esophagus through which water and food pass. The water passes between the branchial bars, over the gill lamellae, and out the opercular flaps, whereas the food is trapped by the gill rakers, lacerated or crushed by the pharyngeal teeth, if present, and swallowed by entering the esophagus. The smaller the food items eaten, the more numerous the gill rakers. Filter feeders have long, slender gill rakers.

Pharynx construction reflects the diet of the fish to a fair degree because it is the site of food retention prior to swallowing. Carnivorous fishes have an array of sharp gill rakers as well as teeth on the tongue, palatine bones, and vomers which aid in retaining the prey and forcing it into the esophagus. Several fishes, including the cyprinids which lack teeth on their jaws, have pharyngeal teeth on the fifth ceratobranchials that bite against a pharyngeal pad on the roof of the pharynx. Many other fishes, such as cichlids and wrasses, have both upper and lower pharyngeal teeth that serve as either retention or crushing devices. The teeth on these bones which line the pharynx are replaced frequently, in the same manner as are the teeth on the jaws (Evans and Deubler 1955). Pharyngeal teeth are charac-

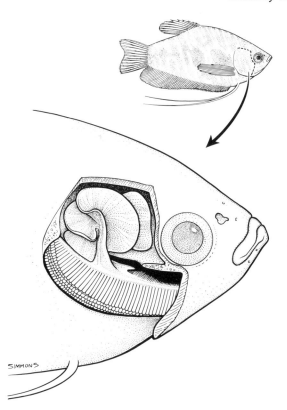

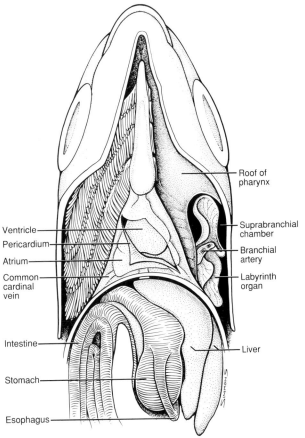

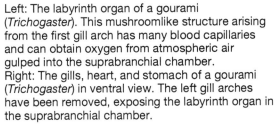

Left: The labyrinth organ of a gourami
(*Trichogaster*). This mushroomlike structure arising
from the first gill arch has many blood capillaries
and can obtain oxygen from atmospheric air
gulped into the suprabranchial chamber.
Right: The gills, heart, and stomach of a gourami
(*Trichogaster*) in ventral view. The left gill arches
have been removed, exposing the labyrinth organ in
the suprabranchial chamber.

teristic for each species and are frequently used
in keys or descriptions.

The digestive tract: In general, carnivorous
fishes have a short gut and herbivorous fishes
have a very long one. In the stone-roller minnow
(*Campostomer*), an algae feeder, the intestine is
six to nine times the length of the body and has
loops that encircle the gas bladder.

The esophagus is short, and in many fishes it
has a duct leaving its dorsal surface that can pass
air into the gas bladder. Such fishes that gulp air
are called physostomous, in contrast to those
with no connection from esophagus to gas bladder, called physoclistous. Often the stomach is
simply a widening of the gut tube with no discern-
ible chamber, as in the swordtail, but it can be
pouchlike and distensible, as in cichlids such as

the oscar. The spleen, which may be triangular or
oval, is usually attached along the stomach or at
the fundus.

At the junction of the stomach and the duode-
num in some species there are pyloric cecae of
variable number and size. There may be only one
or two, or they may number in the hundreds.
Pyloric cecae serve a digestive function and are
capable of reflux emptying.

The small intestine is indistinguishable externally from the large intestine in teleost fishes. If
the intestine opens to the outside, as in the
swordtail, an anus is formed. If it opens into a
cloacal chamber that also receives urinary and
genital products, then the external opening is
called the vent.

The liver is relatively large in fishes and
frequently incorporates the pancreas in whole
or in part. A gallbladder is usually present, and
the bile may be green, brown, or yellow.

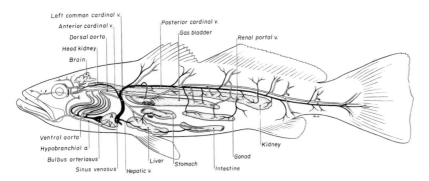

Above: A schematic representation of the circulatory system. Only venous blood (in black vessels) enters and passes through the heart to be aerated in the gills. The atrium of the heart (A) receives blood from the sinus venosus and passes it into the ventricle (V). Below: The heart and gills of a spotted sea-trout (*Cynoscion nebulosus*) in left lateral view. A: First branchial arch connecting ventral to dorsal aorta. B: Cross-section of a gill arch. Note shunt connections between afferent and efferent branchial arteries which allow oxygenated blood to bypass the capillary bed in the gill and continue to the dorsal aorta. C: A schematic gill filament, with several lamellae indicated as cross-connections. Each lamella consists of a platelike capillary plexus between the afferent and efferent filament artery.

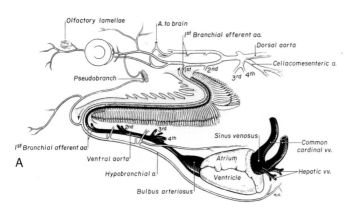

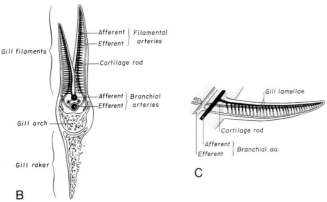

The Respiratory System

Oxygen is carried to the gills in the water current that enters the mouth and leaves through the opercular flaps. The rate at which water is passed through is dependent upon the muscles of the pharynx and operculum or upon the speed of swimming. There is also some oxygen exchange through the skin, pharynx, and gut. Loaches gulp air and pass it through their gut for respiratory purposes. Ganoid fishes have a saccular gas bladder, and lungfish have paired saccular bladders which enable them to cope with adverse environmental conditions.

Several fishes, such as the anabantids (gourami, *Betta*), can utilize atmospheric oxygen by virtue of a labyrinth organ in the pharynx. The labyrinth is an expansion of the first gill arch, with an extensive vascular plexus that projects dorsally into an air-filled chamber of the pharynx. Blood diverted from the gill filaments is passed through this plexus to be oxygenated and then flows into the dorsal aorta.

The Circulatory System

Blood cells are formed in the kidney, liver, and spleen. The erythrocytes are oval and nucleated. Some arctic fishes with low oxygen requirements lack hemoglobin and thus have clear blood. Leukocytes may constitute 10 percent of the blood. Lymphatic vessels are present in teleosts, but there are no lymph nodes. Capillary networks or retia are present in the gas bladder for the introduction or removal of gas (oxygen, nitrogen). Within the red muscle strip of the lateral body wall, the capillary network functions for heat exchange; fishes such as tuna are capable of raising their body temperature 4 to 6 degrees F (approximately 2 to 3 degrees C) above ambient temperatures. In the oviduct and ovarian wall of live-bearing fishes, there are networks for nutrient exchange which can be considered placentae.

Fishes have a one-way circulatory pathway through the heart. All venous blood from the organs, body wall, and fins returns to the heart via superficial and deep veins. It is passed through the four chambers of the heart, without aeration, into the ventral aorta located in the throat. (For a discussion of circulation in fishes, see Satchell 1991.)

Aeration: Aeration is a function of the gills that requires mechanical activity to pass a stream of water over the gill surfaces and out of the opercular opening. Most fishes accomplish this by the use of numerous muscles which open and close the mouth, dilate and constrict the pharynx, and open and close the opercular flaps. The structure of the gill, with its well-vascularized and numerous lamellae, has secondary pathways via shunts that allow the blood to pass from the venous side (afferent branchial artery) to the arterial side (efferent branchial artery) without going through the capillary bed. Such shunts would be used when the oxygen tension is high in an oxygen-rich environment and the demand in the tissues is low due to inactivity. The efferent branchial arteries converge to form a dorsal aorta which courses caudally beneath the vertebral column, giving branches to the kidney, viscera, and body wall along the way.

The heart: The heart lies in a pericardial cavity that is completely separated from the body cavity. Its position is in the throat, ventral to the gills. The heart wall is supplied by hypobranchial vessels from the gill rather than from the dorsal aorta. Likewise, the pseudobranch also receives a special blood supply from the efferent branchial artery.

The heart has four chambers: the sinus venosus, atrium, ventricle, and bulbus arteriosus. The sinus venosus is a thin-walled transverse chamber extending across the dorsal aspect of the pericardial cavity. It receives the common cardinal veins (ducts of Cuvier), which drain the head and body wall from each side, and the hepatic veins, which drain the liver. Backflow of blood is prevented by the sinoatrial valve, which is a fold of the sinus wall extending into the atrium.

The atrium is the largest and most expansible

chamber of the heart. Because of its thin wall and venous blood within, it appears dark. It extends across the pericardial cavity dorsal to the ventricle and balloons around the other chambers.

The ventricle is the thickest walled chamber of the heart and therefore appears white even when filled with blood. It is pyramidal or cone-shaped, with its apex directed anteriorly. It exits into the bulbus of the heart.

The bulbus arteriosus is almost spherical and smaller than the ventricle. It is an intrapericardial portion of the ventral aorta and is rich in elastic fibers. It serves as an elastic relief chamber for the emptying of the ventricle, but is capable of contraction. In sharks, gars, and some other fishes, this chamber is elongated, has a series of three or more valves, and is called a conus arteriosus. It is said that in primitive bony fishes such as shad and bonefish, elements of both the conus and the bulbus are present.

The ventral aorta is the outflow path to the gills. Bony fishes usually have four gill units on each side, and each receives a large branch from the ventral aorta called an afferent branchial artery. This vessel forms capillary nets in the gill lamellae, which then drain into an efferent branchial artery.

The dorsal aorta is formed by the joining of right and left efferent branchial arteries above the gills. It passes caudally beneath the vertebral column, giving rise along the way to paired segmental intercostal, renal, and spinal arteries as well as to unpaired arteries to the gas bladder and viscera. Posteriorly, where the paired ribs fuse to form hemal arches, the dorsal aorta enters the hemal arch and passes into the tail. The caudal vein also lies within the hemal arches beneath the aorta.

The Gas Bladder

The gas bladder, also called the air bladder or swim bladder, is composed of one or two chambers in tandem connected by a short duct. A single-chambered gas bladder may have paired lobes at the cranial end, or the entire bladder may be partially divided longitudinally. In some members of the herring family, the gas bladder may have many diverticulae at the cranial end. Cyprinid fishes have a two-lobed tandem bladder with a sphincter in between. The caudal chamber

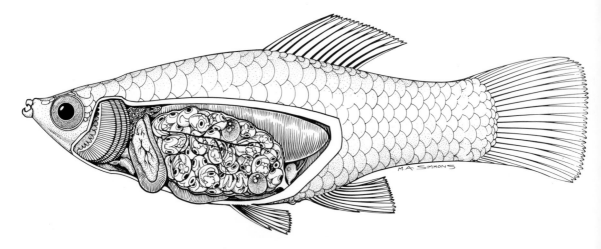

A female swordtail (*Xiphophorus helleri*) with sixty-eight young in late gestation. The ovaries are enclosed in a transparent, vascularized mesovarium. Note that only two mature eggs failed to develop. Fish that give birth to living young, as do most species in the family Poeciliidae, are called viviparous.

is connected with the esophagus via a pneumatic duct, enabling the fish to gulp air and pass it into the bladder. Such fishes with an open connection are spoken of as physostomous fishes; without such a connection, they are called physoclistous. The gas bladder functions primarily as a hydrostatic organ adjusted for neutral buoyancy. A vascular rete, referred to as the "gas gland," secretes oxygen or nitrogen into the bladder (or removes it) so as to allow the fish to remain at a given level. The gas bladder also serves in phonation, audition, and in some fishes such as gars, for accessory respiration.

The Reproductive System

Reproductive specializations in fishes are numerous and often uniquely suited for maintaining maximum survival under adverse environmental conditions. In general, those fishes that congregate in large numbers to spawn (like the cod, eel, and herring) have few if any structural modifications and little if any sexual dimorphism. Fishes that court and pair or raise their young in high-risk habitats have evolved various mechanisms for ensuring fertilization and protecting their eggs. Males and females can differ in size, shape, and color. Sex reversal is common in some (wrasses), and dramatic changes can take place in color, pattern, and size. If a fish is first a functional male, it is said to be protandrous; if first a female, protogynous. There are hermaphroditic, or self-

fertilizing, fishes that have functional ovaries and testes, such as *Rivulus marmoratus* in Florida. The sailfin molly (*Poecilia latipinna*) is gynogenetic: Although there is male courtship and insemination, the sperm only activates the egg, and no fusion of pronuclei takes place.

The ovaries and oviducts are usually paired, although in several species there are examples (such as the swordtail) of fused ovaries with single oviducts. The testes of the male may be paired or partially fused, and sperm are produced in tremendous numbers. Sperm can be released as a cloud over the spawning grounds, or, as in live-bearing fishes, small unencapsulated packets (spermatophores or sperm balls) are inserted into the cloaca of a female via a copulatory organ, the gonopodium.

The gonopodium of poeciliids consists of elongated fin rays 3, 4, and 5, which begin to look different at about fourteen weeks of age in the guppy or eighteen weeks in the swordtail. Accompanying the modifications of the anal fin rays are enlargements of the fin supports and their associated muscles. The fan of muscles on the interhemal spines provides great mobility for the gonopodium, which bends forward and to the side to transfer a sperm mass into the oviduct of the female. Several of the genera in this family crossbreed, and all of them are quite polymorphic, so that many hybrids and varieties of poeciliids have been produced. A variety of terminal hooks and claws on the gonopodium have been shown to be essential for the proper functioning of the mechanism (Rosen and Gordon 1953).

Eggs may be large and few, as in the tilapia and sea catfish, or small and numerous (several million), as in the sturgeon. There are eggs that float (pelagic) and eggs that sink (demersal), eggs that stick (adhesive), and eggs that are carried around in the mouth or on the body of the parent. Some fish eggs hatch in twenty-four hours (*Amphiprion,* the clown anemone fish), three days (*Danio,* the zebrafish), thirty days (*Fundulus,* the killifish), or longer. Eggs may be retained and hatched internally so that live young are produced; these are the viviparous fishes (see Tavolga 1949; Wourms 1981).

About thirteen families (122 genera, 510 species) of teleosts are viviparous (Breeder and Rosen 1966). They may be lecithotropic, in which the young are dependent upon yolk, or matrotrophic, in which the female provides a continuous supply of nutrients, or a combination of both. The eggs in live-bearing fishes either remain within an ovarian cavity (ovisac) or remain within the ovarian follicles of the ovary. In the swordtail (*Xipho-*

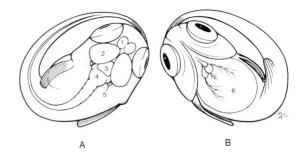

Dorsal (A) and ventral (B) views of a prenatal swordtail (*Xiphophorus helleri*) removed from an ovarian follicle (see previous figure).
1: forebrain; 2: optic lobe of midbrain;
3: cerebellum; 4: hindbrain; 5: otolith;
6: pericardial sac. Note the large brain and the heart with its pericardial sac vasculature that serves as a placenta.

phorus), the paired ovaries fuse into one which has a central lumen continuous with the oviduct. The young develop within ovarian follicles on the inner surface of the fused ovaries, and they derive part of their nutrition from a placenta formed by their pericardial sac.

Reproductive tract of a pregnant swordtail (*Xiphophorus helleri*) in left lateral view. The mesovarium (1) has been reflected and a portion of the left ovary removed to show the cavity between the ovaries and the young within ovarian follicles. (2) Eggs are fertilized within the ovary by sperm deposited in the oviduct (4) via the gonopodium. The young develop within the ovary and utilize their egg yolk and a pericardial placenta (3) for nutrition. When the fish are ready to be born, they emerge into the central lumen and pass through the oviduct and genital pore. There is no cloaca in this species, since the intestine (5) opens separately.

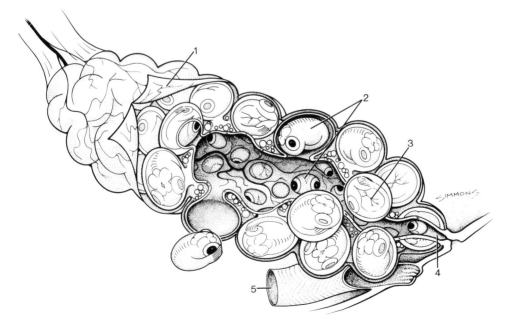

The most common live-bearing aquarium fishes are in the family Poeciliidae, which includes about 136 species fishes in twenty genera now found throughout the world, although all were originally from Mexico and Central and South America. An interesting exception to the live-bearing mode is seen in *Tomeurus,* a monotypic genus with two species from South America. Although this fish normally lays eggs, it has internal fertilization (fin rays 6 to 9 of the anal fin are greatly modified to form a distensible sac when brought forward), and on occasion the eggs are retained and the young are born alive. This latter case is called facultative viviparity.

When two or more broods of embryos of different ontogenetic stages develop simultaneously within the ovisac, it is called superfetation. Thus there may be oocytes maturing and being fertilized within the follicles of the ovary while others are partially developed or even being born. Most poeciliids can store sperm for several months, so the presence of the male is not necessary for all broods. Gestation usually lasts about a month. Eggs within the ovarian follicles of the swordtail are about 2 millimeters (0.08 inches) in diameter and appear to have almost enough yolk for their needs. The dry-weight increase of the swordtail embryo is only about 30 percent during gestation, which is a low level of maternal–fetal transport. The swordtail is considered to be an unspecialized matrotrophe.

At the other extreme, the dry-weight increase of the embryo in *Poeciliopsis turneri* is reported to be 1,840 percent and that of *Heterandria formosa,* 3,900 percent. The latter two fishes are considered to be specialized matrotrophes with a follicular placenta and superfetation. In the four-eyed fish (*Anableps*), the postfertilization weight has been reported (Knight et al. 1985) to increase 298,000 percent. Placental transfer in this fish is accomplished by apposition of the maternal follicular epithelium (within follicular pits) to absorptive surface cells (vascular bulbs) of the pericardial trophoderm. The pericardial trophoderm is an extension of the somatopleure, which replaces the yolk sac during early gestation. The follicular placenta of *Anableps* is probably the most efficient maternal–embryonic nutrient transfer system in teleost fishes.

The Urinary System

The kidneys are elongate structures that lie below the vertebral column and may extend from the head to the end of the abdominal cavity. They are sometimes expanded cranially and/or caudally, as in the oscar, or at midbody, as in the goldfish. The kidneys in adult fishes are mesonephric in structure and function for excretion as well as blood formation. Gobies retain a functional pronephric kidney, but in most fishes there is only a remnant on each side of the embryonic pronephric kidney at the level of the first five vertebrae. This head kidney functions as a lymphoid organ. The remainder of the kidney is mesonephric and drains via a mesonephric duct into a bladder or into a cloaca. Several fishes, including the swordtail, have urinary bladders.

For more information on the biology and anatomy of fishes see Bond (1979), Harder (1975), Love (1970, 1980), and Moyle and Cech (1982).

Selected References

Bond, C. E. 1979. *Biology of Fishes*. Philadelphia: W.B. Saunders.

Breder, C. M., Jr., and Rosen, D.E. 1966. *Modes of Reproduction in Fishes*. New York: Garden City Press.

Evans, H. E., and Deubler, E. 1955. Pharyngeal tooth replacement in *Semotilus atromaculatus* and *Clinostomus elongatus,* two species of cyprinid fishes. *Copeia* 1: 31–41.

Gordon, M. 1927. The genetics of a viviparous top minnow *Platypoecilus:* the inheritance of two kinds of melanophores. *Genetics* 12: 253–83.

———. 1932. The scientific value of small aquarium fishes. *New York Zool. Soc. Bull.*: 1–8.

Harder, W. 1975. *Anatomy of Fishes*. Part I. Text. Part II. Figures. Stuttgart: E. Schweizerbart'sche Verlag.

Kapoor, F. M.; Evans, H. E.; and Pevzner, R. 1975. The gustatory system in fish. *Adv. Marine Biol.* 13: 53–108.

Knight, F. M.; Lombardi, J.; Wourms, J.P.; et al. 1985. Follicular placenta and embryonic growth of the viviparous four-eyed fish (*Anableps*). *J. Morphol.* 185: 131–42.

Love, R. M. 1970. *The Chemical Biology of Fishes*. New York: Academic Press.

————. 1980. *The Chemical Biology of Fishes.* Vol. 2. *Advances, 1968–1977.* New York: Academic Press.

Moyle, P. B., and Cech, J., Jr. 1982. *Fishes: An Introduction to Ichthyology.* New York: Prentice Hall.

Nelson, J. S. 1984. *Fishes of the World.* 2nd ed. New York: John Wiley & Sons.

Rosen, D. E., and Gordon, M. 1953. Functional anatomy and evolution of male genitalia in poeciliid fishes. *Sci. Contrib. N.Y. Zool. Soc.* 38(1): 1–47.

Satchell, G. H. 1991. *Physiology and Form of Fish Circulation.* Cambridge: Cambridge University Press.

Tavolga, W. N. 1949. Embrynoic development of the platyfish (*Platypoecillus*), the swordtail (*Xiphophorus*), and their hybrids. *Bull. Amer. Mus. Nat. Hist.* 94(4): 161–230.

Vielkind, J., and Vielkind, U. 1982. Melanoma formation in fish of the genus *Xiphophorus:* a genetically-based disorder in the determination and differentiation of a specific pigment cell. *Can. J. Genet. Cytol.* 24: 133–49.

Wourms, J. P. 1981. Viviparity: the maternal–fetal relationship in fishes. *Amer. Zool.* 21: 473–515.

Fish Physiology

Robert E. Reinert

In the broadest sense, an understanding of fish physiology includes an understanding, in chemical and physical terms, of all the mechanisms that operate in fish. Since there are more species of fishes than of all the other species of vertebrates put together, and because of the extremely wide variety of habitats fishes occupy, there are often large differences in the physiological strategies they use to live in particular environments. Even if we only consider the physiology of one species, the task of trying to explain all the factors involved in how that species functioned would not only be beyond the scope of this chapter, but beyond that of any reasonably sized textbook. Therefore, this chapter is only an introduction to a few of the basic physiological functions of fishes. Included will be the senses, respiration, osmoregulation, and the stress response. It is our hope that this introduction to fish physiology will provide some initial insight into how fishes function in their environment and encourage the reader to learn more about these intriguing animals.

The Sensory Systems

In fishes, the sense organs provide the only input from the outside world to the nervous system. Thus, the reactions of fishes are largely the results of responses to external stimuli that have been transduced to electrical impulses by the various sensory receptors and relayed to the brain. Most fishes possess traditional vertebrate sense organs that provide for sight, hearing, touch, taste, and smell. They also possess the lateral line system, which is unique to fishes and the aquatic stages of amphibians. In the following pages, we will discuss some of the basic principles involved in the operation of these senses and their importance to the overall well-being of fishes.

Vision

Morphology: In many ways the fish eye is structurally similar to the eye of terrestrial vertebrates. For example, both have a cornea, an iris, a lens, and a retina that contains rods or rods and cones. However, because air and water have different effects on the properties of light, a number of features of the fish eye have been modified so it can function at a high level of efficiency in water. One principal difference between the eye of most terrestrial vertebrates and most fishes is the shape of the lens. Anyone who has ever looked at an underwater object while swimming knows that it is impossible to see the object clearly unless you are wearing a face mask. This is because the greater the density difference between two mediums, the greater will be the angle of refraction as the light passes from one to the other. When light strikes our eye in air, the majority of the refraction occurs at the air–cornea inter-

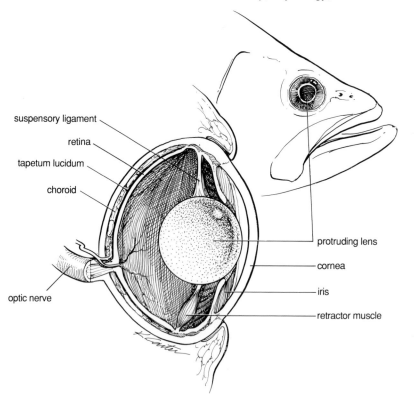

suspensory ligament

retina

tapetum lucidum

choroid

optic nerve

protruding lens

cornea

iris

retractor muscle

The spherical lens of most fish eyes focuses by moving within the eyeball.

face. The primary purpose of our lens is to fine-tune this refraction and focus the light on the retina. In water, however, the density of water and our eye are about the same. Consequently, there is little refraction as the light passes into our eye, and because our lens is somewhat flattened, it is not capable of refracting the light enough to focus it on the retina. The layer of air in a face mask bends the light enough so our lens can focus it. To compensate for the lack of refraction by the cornea, the lens in most fish eyes is spherical. This shape and the high density of the lens result in a refractive index of about 1.67. This is the highest refractive index of any vertebrate lens and is sufficient to bend the light so it can be focused on the retina.

The shape and relative rigidity of most fishes' lenses also necessitates that fish have a method of focusing (accommodation) that is different than that of most terrestrial vertebrates. For example, humans focus light on the retina by changing the shape of the lens. Most fishes, however, do not change the shape of their lens because they must maintain the optimal refractive index. Instead,

they focus by changing the position of the lens in much the same manner as a camera is focused. The relaxed position in most teleost (bony) fishes is when the lens is in the forward position. In this position, near objects in front of the fish are in focus. Accommodation to more distant objects in front of the fish is accomplished by moving the lens back toward the retina with the retractor lentis muscle. In many sharks the lens is closer to the retina in the relaxed position. These fishes accommodate to close objects by moving the lens forward with the protractor lentis muscle.

Another difference between the teleost eye and that of most terrestrial vertebrates is that the lens of most teleost fishes protrudes through the pupil. Along with the ellipsoid shape of the retina, this gives most fishes an extremely wide angle of vision that includes the area in front and continues laterally to almost directly behind them. Also, when fish swim, the side-to-side head movement coordinated with eye movements tends to eliminate any blind spots behind them.

The structure of the eye in some elasmobranchs (sharks, skates, and rays) differs in several ways from that of the teleost eye. The elasmobranch lens is often more flattened and does not extend through the pupil. Also, unlike the fixed pupil of most teleosts, many elasmobranchs can control the amount of light entering their eye by constricting or dilating their pupil in a manner similar to that of terrestrial vertebrates. The light-adapted pupil has been shown to take many shapes, from that of a pinhole in the blacktip shark (*Carcharhinus limatus*) to a horizonal slit in the bonnet-head shark (*Sphyrna tiburo*). Also, although most teleosts lack any type of protective covering over the cornea, many sharks have a nictitating membrane. This membrane probably protects the eye when a shark is feeding, because

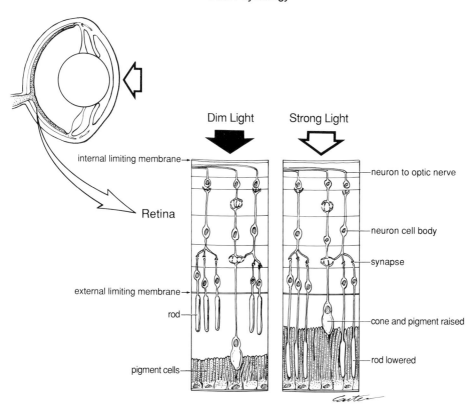

Dim Light   Strong Light

internal limiting membrane →

Retina

neuron to optic nerve

neuron cell body

synapse

external limiting membrane →

rod

cone and pigment raised

rod lowered

pigment cells

Movement of rods, cones, and pigment in the retina of a teleost fish. In light adaptation, the rods are moved away from the light and are protected by forward movement of pigment.

it closes over the eye just before the shark strikes its prey.

The retina of a fish is organized in a manner similar to those of other vertebrates. Closest to the light is a clear layer of nerve cells and nerve fibers. Just behind this layer are the photosensitive rods and cones. The outermost layer of the retina is the pigmented epithelium. In many teleosts, there is a reflective mirrorlike area, the tapetum lucidum, interspersed among the cells of the pigmented epithelium. Guanine crystals are the reflective substance in the tapetum lucidum of most fishes. Some teleosts, however, have other reflective materials. A yellow "melanoid" substance has been reported to occur in gars (Lepisosteidae) and catfishes (Ictaluridae), and a pteridine appears to be the reflecting substance in gizzard shad (*Dorosoma cepedianum*). The tapetum increases the sensitivity of the eye in low light by reflecting light back over the rods. Similar materials found in the eyes of other animals, such as cats and

deer, are the cause of the so-called "glassy-eyed" look when light shines into their eyes.

In fishes, as in other vertebrates, rods appear to be primarily for vision in dim light and cones primarily for color vision in more brightly lit situations. Rods show a higher degree of convergence in their neural connections (a number of rods may be connected to a single nerve fiber) than cones and consequently respond to lower levels of light.

The number of rods and cones and their distribution in the retina vary with different species of fishes. Generally, those that are active at twilight (crepuscular) or live in low-light environments have a higher ratio of rods to cones than those that live in more brightly lit environments. For example, the crepuscular eel-pout (*Lota lota*), has a rod:cone ratio of about 238:1, whereas fishes such as the mackerel sharks (Lamnidae), which are active in a more highly lit environment, have a rod:cone ratio of 6:1. Because most deep-sea fishes live in an environment with very little light, their retinae contain either only rods or rods with a very sparse number of cones.

The fovea is a depression in the retina containing closely packed cone cells for increased visual acuity. Although it is common in higher vertebrates, it is rare in fishes but has been reported for some species. For example, three species of serranid basses have been shown to have well-developed depressions containing high concentrations of cones. The exact position of these areas on the retina seems to be related to the horizontal or vertical feeding habits of the fish.

Light/dark adaptation: An interesting feature of the teleost eye is the manner in which these fishes control the amount of light striking their rods and cones. Humans and many other higher vertebrates control the amount of light entering their eye by contracting or dilating their pupil. However, because their lens extends through the pupil, most teleosts have adapted another method, which involves synchronized movements of the rods, cones, and pigment granules in the pigmented epithelium. Under conditions of bright light, the pigment granules migrate toward the outer segments of the rods and cones. At the same time, the rods move back into the pigment and the cones move toward the light. This arrangement protects the rods from the effects of bright light by enclosing them in the pigments and at the same time exposes the cones to the light. In low light, the pigment granules move away from the rods and cones. At the same time, the rods move toward the light and the cones move away from it. The time required for light and dark adaptation in fishes is considerably longer than the contraction or dilation of the pupil in higher vertebrates. For instance, in young salmon, light and dark adaptation take about twenty-five minutes and one hour, respectively.

Visual pigments: The light environment of fishes is much more varied than that of most terrestrial vertebrates. This is due to the light-scattering and absorption properties of water and the materials in water. Because of these large differences in both quality and quantity of light in their environment, fishes have evolved the widest range of visual pigments of any of the vertebrates. The two basic groups of visual pigments in fishes are rhodopsins and porphyropsins. Classically, rhodopsins are found in marine fishes and porphyropsins in freshwater fishes. However, many freshwater fishes, such as some species of

charcids and cyprinids, have mixtures of both pigments. Some migratory fishes, such as lampreys (Petromyzontidae) and coho salmon (*Oncorhynchus kisutch*), also have both pigments and undergo seasonal changes in their rhodopsin-porphyropsin ratios.

All of the visual pigments studied in fishes have two major components: a lipoprotein moiety called an opsin, and a chromophore consisting of $retinene_1$ (an aldehyde of vitamin A_1) or $retinene_2$ (an aldehyde of vitamin A_2). The rhodopsins, which are similar to the visual pigments found in humans, contain $retinene_1$ and the porphyropsins, $retinene_2$. Within each of these two major groups of visual pigments, differences in the opsins determine which wavelengths of light are maximally absorbed by the pigment. When light is absorbed, the pigment undergoes conformational changes that result in the generation of an electrical impulse. This impulse is transmitted to the brain, where its integration results in the sensation of sight.

There have been a number of attempts to relate the maximum spectral absorption qualities of visual pigments to the different spectra found in various aquatic environments. In many marine teleosts, the visual pigments appear to be those that are most sensitive to the wavelengths that predominate in the environment in which the fishes live. For example, many fishes living in the deeper waters of the ocean, where all but blue light (shorter wavelengths of about 470 to 480 nanometers, or nm) is filtered out, have pigments that maximally absorb these shorter wavelengths. Conversely, many fishes that inhabit shallower, more turbid waters have pigments that are shifted toward the longer yellow-green wavelengths that penetrate these waters. Other fishes have pigments that appear to maximize the contrast between prey and the background light. In some predatory reef fishes, the visual pigments are most sensitive to the wavelengths that predominate at dawn and dusk, when these fishes are most active.

Color vision: One of the best indications that some fishes have color vision is that they have cones, which are the color receptors in higher vertebrates. In order to distinguish differences in wavelength (i.e., color) from differences in brightness, an organism must have receptors that respond to different wavelengths. In 1801 Thomas Young correctly suggested that three basic types of color receptors could account for the sensation of color vision. For example, in humans there are three basic types of cones that have pigments that have their maximal sensitivities at 440 nm (blue), 535 nm (green), and 575 nm (yellow), respectively. In most fishes that have color vision, there are also different types of cones. Goldfish (*Carassius auratus*) have three types of cones that have maximal sensitivities at 455 nm (blue), 535 nm (green), and 620 nm (red). Basically, the theory for color vision is that blue, green, or red light will cause the particular cone that is sensitive to it to respond. The result will be the sensation for that particular color. However, if the color is some combination of these, the cones will respond more or less in proportion to the amount of blue, green, or red in that color, and the sensation will be for that color.

The most conclusive evidence of color vision in fishes comes from conditioned-response tests. These tests involve conditioning a fish to a particular event and then measuring the response. For example, a goldfish can be shown a number of cards that are different hues but the same brightness. Several seconds after being shown a particular hue, the fish is shocked. By monitoring the heartbeat, it can be shown that, often after only a few trials, the fish learns to associate that particular hue with the impending shock. The heart rate declines in anticipation of the shock. This and other types of conditioned-response tests have demonstrated that many fishes do respond to color and often over much the same range as humans. The primary biological advantage suggested for color vision over non-color vision is that it enhances the perception of various targets against a variety of backgrounds.

Hearing and Equilibrium — The Inner Ear

Underwater sound: Although sound obeys the same physical laws in air and water, many quantitative differences result from differences in the compressibility and density of the two media. For example, in both water and air, a sound wave results from the compression of particles that rebound after compression and impart their directional energy to neighboring particles. However, because molecules in water are closer together than in air, the motional energy is transferred from molecule to molecule faster. Therefore, sound travels about 4.8 times faster in water than in air.

The near-field and far-field effects are another factor that plays an important part in underwater acoustics. These terms refer to the ratio of the pressure of the sound waves to particle displacement (movement of the water). In the near-field, the pressure of the sound wave and the particle displacement are out of phase. Particle displacement increases more rapidly than pressure as one moves closer to the sound source. Therefore, in the near-field, small pressure changes result in relatively large changes in particle displacement. In the far-field, particle displacement and pressure have a constant ratio. The point at which the ratio between the two becomes constant is termed the near-field/far-field boundary. Because water is much denser and therefore less compressible than air, the extent of the near-field effect in water is about five times greater than in air. Also, the lower the frequency of the sound source, the greater will be the extent of the particle displacement.

A general rule for determining the extent of the near-field effect is that the near-field/far-field boundary will be about one-sixth the wavelength from the sound source. For example, at a frequency of 20 cycles per second, the near-field will extend about 12.5 meters (41.0 feet) from the sound source. At a frequency of 1,000 cycles per second, it will only extend for about 0.25 meters (0.82 feet). Because of differences in the acoustical characteristics of air and water, terrestrial vertebrates almost always are exposed to sound in the acoustical far-field. Fishes, however, are exposed to both and have evolved the lateral line system, which is especially sensitive to particle displacement caused by low-frequency vibrations.

Most of the sound generated in either air or water is reflected back into that medium at the air–water interface. This means that fishermen in a boat can talk without being detected by the fish, but any sounds such as the banging of a tackle box or the scraping of feet will be transmitted into the water and possibly be detected by the fish. In the aquatic environment, along with compression waves, any movement of water such as fin movement, water currents, the movement of a fishing lure through the water, the bow wave generated when a fish swims, or the scraping of a tackle box on the bottom of a boat is considered an acoustical phenomenon. The two sound-detection systems in fishes are the inner ear and the lateral line system.

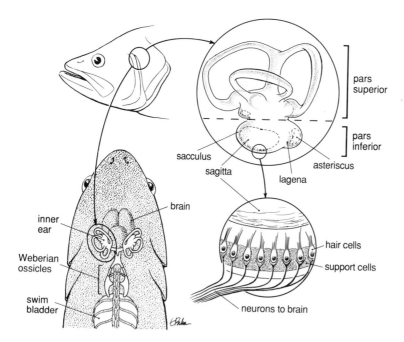

Fish have no outer or middle ear, but do possess an inner ear (above, right) composed of interconnected ducts and chambers with otoliths inside. A cross-sectional magnified view (below, right) shows a portion of an otolith resting on sensory hair cells.

Morphology of the inner ear: Fishes do not have an outer or middle ear; however, they do have an inner ear that is similar in many ways to the inner ear of higher vertebrates. In both groups of animals, the inner ear consists of a membranous system of connecting ducts and chambers. In most teleost fishes, the lower portion (the pars inferior) has two chambers, the sacculus and the lagena. Each of these chambers contains a calcarious ear stone called an otolith. The upper portion (the pars superior) consists of three semicircular canals that extend through the horizontal, vertical, and longitudinal planes and one chamber, the utriculus, which also contains an otolith. Each of the semicircular canals has an ampulla (chamber) that responds to changes in the inertia of the fluid in the semicircular canals. In most of the fishes studied, the primary function of the pars superior seems to be as an equilibrium and gravity detector, and the main function of the pars inferior is sound detection.

Hearing: Because a fish is about the same density as water, sound waves move through its body at about the same amplitude and frequency as they move through the water. If the sound waves moved through all of the fish's body at the same amplitude and frequency as they move through the water, the fish would not be able to detect sound. The otoliths, however, are about three times as dense as most of the other tissues. Therefore, when a sound field strikes an otolith, it is not as easily set in motion as the rest of the tissues. Consequently, the phase and amplitude of the otolith movement are different from that of the rest of the body. In each chamber, the otolith is suspended in fluid and rests on a sensory epithelium (macula) composed of hair cells. From each hair cell an apical bundle of cilia (minute, short hairlike processes) projects into the chamber. The out-of-phase movement of the otolith with regard to the cilia stimulates (bends) the cilia. This causes the hair-cell membrane to change shape. Depending on the direction in which the cilia bend, there is an increase or decrease in the rate of impulses from the sensory axon with which the hair cell makes contact. These changes in firing rate travel to the auditory center of the brain and result in the sensation of hearing.

The swim bladder is important to the hearing of fishes because it contains a different medium than that through which the sound is traveling. The swim bladder is a saclike structure containing gases that is located in the abdominal cavity above the viscera. Because gas in the swim bladder is more compressible than water, sound waves cause the walls of the swim bladder to vibrate. This in turn causes surrounding tissues in the fish to vibrate, which stimulate the otolith organs. It has been shown that deflating the swim bladder decreases hearing sensitivity.

A wide variety of sizes and shapes of swim bladders are found in different species of fishes. For example, in herrings (Clupeidae) and anchovies (Engraulidae), projections of the swim bladder enter the skull and press against the wall of the inner ear. In adult mormyrid fishes (a group of

African freshwater fishes), portions of the swim bladder are separate from the main swim bladder and are enclosed within the skull in contact with the inner ear. There have been few studies concerning the relationship between hearing and the shape, size, proximity to the inner ear, and complexity of the swim bladder in fishes. Until there are more studies concerning these factors, it will be impossible to assess the overall role of this organ in the hearing of fishes.

The Weberian ossicles are a series of three or four small bones derived from the vertebrae and are very important in the hearing of some fishes. These bones couple the swim bladder to the inner ear and are attached to the vertebrae by elastic cartilage that allows them some freedom of movement. The Weberian ossicles are found only in minnows, catfishes, and other fishes belonging to the superorder Ostariophysi. Their primary function appears to be to mechanically transmit vibrations from the swim bladder to the inner ear. It has been shown that removal of one of these bones will decrease the range of hearing.

There is no doubt that there are substantial differences in hearing abilities between different species of fishes. It is difficult, however, to determine what the true hearing abilities of fishes are in natural surroundings. This is primarily due to the acoustical problems associated with most testing procedures. For example, most hearing studies with fishes have been done in small aquaria where it is difficult to separate the effects of the near- and far-field. Also, reflections and alterations of sounds at the air–water interface and by the sides and bottom of the aquarium make it very difficult to generate "pure" sounds. One way to alleviate these problems is to work in a more natural open environment. This approach, however, presents other difficulties, such as the presence of ambient noise and the problems associated with trying to observe fishes from considerable distances.

Fishes with the broadest range and the lowest threshold of hearing are those with Weberian ossicles. In order of decreasing range and increasing threshold, they are followed by fishes with swim bladders. Fishes with the poorest hearing are those without swim bladders, such as the sharks. For example, the upper limit of hearing for most ostariophysian fishes with Weberian ossicles is 6,000 to 13,000 cycles per second (Hertz, or Hz). By comparison, the upper hearing range of young

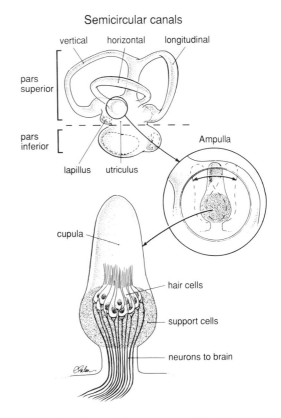

Increasingly enlarged views of one of the three ampullae of the inner ear, which serve as sensory receptors for equilibrium.

adult humans is about 20,000 Hz. The upper limit for fishes with swim bladders but without Weberian ossicles is about 800 to 3,000 Hz. Sharks respond to frequencies of 55 to about 500 Hz. It should be noted, however, that while there does seem to be a general trend in the hearing abilities of fishes within these groups, there are extremely large variations from species to species within and between the groups.

Equilibrium: Anyone who has ever been scuba diving quickly realizes that in an aquatic environment there is a sensation of being weightless in a three-dimensional world. When there is a lack of visual clues, it is very easy to become disoriented and lose all sense of direction. Fishes, however, have a very sensitive system of gravity and directional equilibrium receptors that allow them to sense their position at all times. The otoliths are generally associated with gravity detection in fishes. When the fish changes position relative to gravity, the otolith moves on the macula. This causes the sensory cilia to bend and results in a change in the rate at which impulses from the otolith organ are sent to the brain. In theory, by maintaining the same impulse rate in the otolith organs from both inner ears, a fish would be in balance relative to gravity.

The semicircular canals (one in hagfish, two in lamprey, and three in all other fishes) are the organs of dynamic or spatial equilibrium in fishes. As is the case with higher vertebrates, each of the three canals is aligned in a different plane. The horizontal plane responds to side-to-side motion of the head, the vertical plane to up or down motion of the head, and the longitudinal plane to rotation about a head-to-tail axis. Perhaps a better way to visualize these motions is to think of the yaw, pitch, and roll of a ship. These motions are sensed in the ampullae of the semicircular canals, each of which contains hair cells similar to those on which the otoliths rest. Cilia from a number of hair cells (generally twenty to seventy) are covered with a gelatinous matrix called a cupula, which extends up into the canal. When a fish moves in any direction, there is a corresponding movement of the endolymph fluid in the canal. The movement of the fluid causes the cupula and its associated cilia to bend. This in turn changes the firing rate of the neurons associated with the hair cells. These changes in the spontaneous firing rate are transmitted to the equilibrium centers in the medulla, and the fish makes the appropriate body and eye movements to maintain its position in the water. A fish is constantly reacting to changes in impulses from its gravity and dynamic equilibrium systems. Even at rest, many fishes must continually make adjustments with their fins to maintain their equilibrium because their center of gravity is slightly above the midline through their body. If this type of fish loses the ability to react to changes in these impulses, it will roll over on its back. This phenomenon is often observed by fishermen when they try to return fishes such as largemouth bass (*Micropterus salmoides*) or northern pike (*Esox lucius*) to the water.

The Lateral Line System

The sensory systems discussed thus far—vision, hearing, and equilibrium—are all senses that are familiar to us because of our own experience with them. The lateral line system, however, is found

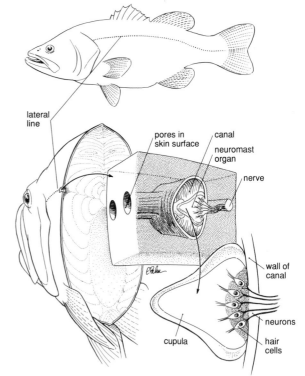

Increasingly enlarged views of the lateral line system, which senses water movement.

only in fishes and the aquatic stages of amphibians, and therefore is harder for us to conceptualize. Perhaps the most descriptive term for this system is the German word *Ferntastsinn,* which means "distant touch." This system responds to any type of water movement, including currents, reflections of the fish's bow wave, and movements of other fishes.

Morphology: The sensory receptors of the lateral line are called neuromast organs and are similar to those of the ampullae organs of the inner ear. Each neuromast organ consists of a number of hair cells enclosed in a cupula. In a manner similar to the auditory and equilibrium hair cells, movement of the cupula and the associated movement of the enclosed cilia cause an increase or decrease in the frequency of the impulses sent to the brain. There are two basic types of lateral line receptors: epidermal organs, which are found in the skin with the cupulae extending into the surrounding water; and canal organs, which are located in a system of canals beneath the skin. All fishes have one or both of these types of receptors. Some of the more primitive species, such as the hagfishes and the lamprey, only have epidermal organs. Most of the more advanced fishes, however, also have well-developed lateral line canals. In many fishes, not only do the lateral line canals extend along the sides of the fishes, but they often branch and extend out along the head. Neuromast organs in these canals are usually arranged in series along the sides of the fish, with their cupulae projecting well up into the canal. Water movement in any direction striking the sides of the fish causes the mucus in the lateral line canal to vibrate, which stimulates the neuromast organs.

Function: There is no doubt that the lateral line system is extremely important to fishes. The main function of this system is the detection of particle displacement that results from low-frequency vibrations or the movement of water. For instance, even without the use of their inner ear, most fishes still respond to low-frequency vibrations below 200 Hz. However, when the nerves innervating the lateral line are cut, these responses are lost. The lateral line is very important in the detection of prey and predators. In water where the visual field is poor, it is probably the lateral line system that first detects the presence of a fishing lure moving through the water. Also, a blinded fish can avoid solid objects due to the sensitivity of the neuromast organs to changes in the bow waves reflected back to the fish. Fish in an aquarium probably use their lateral line system to sense the location of the aquarium walls. Along with vision, the lateral line also appears to play an important part in the schooling behavior of some fishes. For instance, pollock (*Pollachius virens*) with their eyes covered maintained their position in a school with nonblinded fish. This schooling ability was lost when the lateral line nerves were cut.

An interesting generality concerning fish behavior and the lateral line is that the more active the species, the greater is the percentage of canal organs compared to epidermal organs. The explanation is that the skin between the canal and the water acts as a buffer between the water continuously passing over the fish and the neuromast organs, and thereby lessens the problem of "environmental noise."

Chemoreception — Olfaction and Taste

The sensory systems discussed thus far use two types of sensory receptors. Photoreceptors transduce energy from the electromagnetic spectrum into nervous impulses responsible for sight, and mechanoreceptors transduce the movements of water into electrical impulses responsible for the "distance touch" sense of the lateral line and the hearing and balance functions of the inner ear. The sensory receptor organs for the next two senses, olfaction and taste, are categorized as chemoreceptors. They transduce chemical phenomena into nervous impulses responsible for the sensations of olfaction and taste.

Because chemoreception in fish takes place in a medium where the molecules are frequently dissolved in water, it is often difficult to distinguish whether olfaction or taste is responsible for a particular reaction. We do know, however, that fishes do have distinct olfactory and taste organs that are similar in morphology to those of terrestrial vertebrates. Olfaction in fishes is primarily important for detecting substances at a distance, and taste, in most fishes, is considered a contact sense for testing the palatability of substances. If one considers that the molecules responsible for the sensations of olfaction and taste in terrestrial vertebrates must enter a mucous layer before they contact the sensory organs, chemoreception in both fishes and terrestrial vertebrates can be considered an aquatic phenomenon.

Olfactory organ morphology and function: In most fishes, olfactory organs are more sensitive than organs for taste. Thus, they are more important for sensing dilute quantities of substances at a distance. The olfactory organs of fishes are similar to those of humans and other higher vertebrates in that they are located in the same general area and have the same innervation. However, unlike humans, only rarely in fishes does the olfactory organ open into the mouth.

Among fishes there are a number of variations in the morphology of the olfactory organ. The simplest form is a single median nostril with just one opening, as found in the primitive lampreys and hagfishes. Sharks and rays have paired olfactory organs located on the ventral side of the snout. The opening of each organ is divided into two parts by a fold of skin. Most teleosts have paired olfactory organs located on the snout above the mouth. In most instances, each olfactory organ has a separate inlet and outlet for water, which is passed through the olfactory organ in three ways: 1) breathing movements compress an olfactory pouch and force water in and out of a single opening in rhythm with the respi-

Fish Anatomy, Physiology, and Nutrition

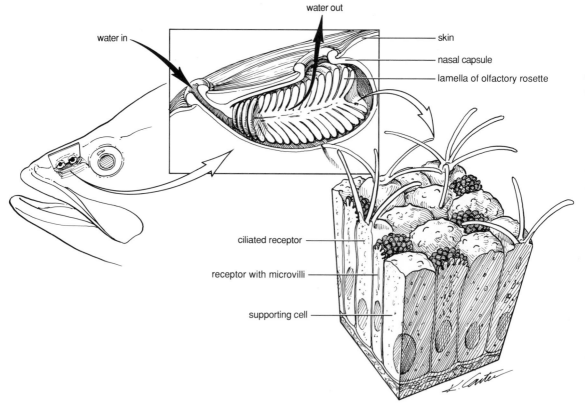

water out

water in

skin

nasal capsule

lamella of olfactory rosette

ciliated receptor

receptor with microvilli

supporting cell

Increasingly enlarged views of the olfactory organ of a fish. As water passes through the organ, odorants contact the surface of sensory cells.

ratory movements (as in the three-spine stickle-back, *Gasterosteus aculeatus*); 2) cilia, or breathing movements, or both, of the fish cause water to circulate through the anterior opening and out the posterior opening (as in the yellow perch, *Perca flavescens*); 3) water is deflected in and out of the olfactory organ by the fish's movement through the water (as in the northern pike).

The shape of the olfactory sac in fishes varies considerably from species to species. The three basic shapes are: round, which is generally associated with fishes that are primarily sight feeders; oblong, which is generally associated with fishes that use both vision and olfaction to locate food; and elongate, which is generally associated with fishes that have poor vision and a highly developed sense of smell. There is also considerable variation in the arrangement of the sensory epithelium that lines the olfactory sac. In most fishes, the epithelium is raised from the floor of the sac and forms a series of rosettelike folds, which greatly increase the surface area. The number and arrangement of these folds also varies a great

deal among species, and together with the size and shape of the olfactory sac generally correlates with the importance of olfaction to the fish. For example, species of puffers (Tetraodontidae), which are highly visual reef fishes that live in very clear water, have no olfactory sacs; rainbow trout (*Oncorhynchus mykiss*), which rely on both vision and olfaction, have an oblong-shaped olfactory sac with about eighteen folds; and channel catfish (*Ictalurus punctatus*), which have a highly developed sense of smell, have an elongate-shaped olfactory sac with about 142 folds. Because the number of lamellae in the olfactory sacs of some fishes has been shown to increase with growth, it has been suggested, but not proven, that as fishes grow older, their sense of smell becomes more acute.

The olfactory sensory receptor cells in fishes are bipolar neurons with cylindrical dendrites that terminate at the surface of the epithelium. Olfaction is the only sense in which the receptor cell is a direct extension of its respective cranial nerve and not connected to it by one or more synapses.

There appear to be two general categories of receptor cells, which are distinguished by the form of the distal free end of the dendrite: cells having cilia and cells having microscopic projections (microvilli) on their surface. Although it has not been conclusively proven, it is generally assumed that olfactory receptor sites are located on cilia or microvilli that extend above the surface of the epithelium. There appear to be several basic types of arrangements of the sensory cells on the olfactory epithelium. These range from sensory cells that cover the entire epithelium, except for the dorsal part of the folds, to small isolated islands of sensory cells arranged somewhat like taste buds.

We do not know how a fish or any other animal transduces information about an organic molecule into the sensation of smell. The most accepted theory is that the cilia or microvilli of the receptor cell have different receptor sites that respond to the steric fit of certain shaped molecules. Odors appear to be correlated with the steric conformation of the molecules. For example, in humans seven basic odors have been postulated: camphoraceous, musky, ethereal, floral, pungent, putrid, and pepperminty. It has been observed that substances with similar molecular configurations have similar odors. According to the theory, molecules with intermediate shapes produce intermediate odor sensations by fitting, and thereby stimulating, several different receptor sites. This would explain how a relatively few different types of receptor sites could be responsible for many different odor sensations.

Sensitivity to pure compounds: Both electrophysiological and behavioral techniques have been used to measure the sensitivity of olfactory organs of many different fishes to a wide variety of materials. These tests have shown there is a wide range of sensitivities among fishes and between different organic materials. Fishes with highly sensitive olfactory organs are termed "macrosmatic," and those with less highly developed systems are termed "microsmatic." Good examples of microsmatic fishes are the minnow (*Phoxinus phoxinus*) and rainbow trout. Both can detect β-phenylethyl alcohol at concentrations in the low parts per trillion. This range of sensitivities is about the same as that of humans. Compared to many other terrestrial animals, humans are not considered to have a highly developed sense of smell. However, if one considers that 2 1/2 ounces

(about 75 grams) of a material dissolved in 1 million 20,000-gallon railroad tank cars of water is equivalent to about 1 part per trillion, it is easy to see that olfaction is a highly sensitive detection system in these fishes and in humans. The American eel (*Anguilla rostrata*) can truly be classified macrosmatic. Its lowest threshold for the above-mentioned chemical is about 80 to 100 times lower than that of the minnow or rainbow trout. This degree of olfactory sensitivity is about the same as that of macrosmatic terrestrial animals such as dogs.

Amino acids are a class of compounds that have been studied a great deal in relation to their effect on the olfactory system of fishes. Because they are the building blocks for protein, it is not surprising that a number of researchers have tried to find which ones or what combination might prove to be an attractant to fishes. For instance, in field tests where amino acids were pumped into water and their attractiveness to winter flounder (*Pseudopleuronectes americanus*), mummichog (*Fundulus heteroclitus*), and Atlantic silversides (*Menidia menidia*) was measured, a number of amino acids were found to be effective attractants. Although there was a good deal of overlap in preferences among species, different species did prefer different amino acids. In laboratory tests in which different amino acids were introduced into large tanks containing carp (*Cyprinus carpio*), the most positive responses were to the amino acids valine and lysine. In general, electrophysiological and behavioral tests with a variety of fishes suggest that of the classes of compounds tested, amino acids appear to have the greatest stimulatory effect. Other pure substances that evoked the feeding response were ammonia, lactic acid, creatinine, and glutamic acid.

Repellents: An interesting finding concerning amino acids was that the odor from biologists' hands rinsed in water running down a fish ladder acted as a repellent to coho salmon. The most active ingredient from the hand rinsings was found to be the amino acid L-serine, which is a common amino acid in mammalian skin. It has been suggested that the avoidance response to L-serine has evolved as a method for alerting migrating salmon to the presence of mammalian predators such as sea lions and bears. Based on this information, a number of products have been developed that are supposed to mask the human scent when sprayed on a fishing lure. In addition, some are advertised to contain compounds that act as fish attractants.

It is interesting to note that the electrical response of the olfactory epithelium is about the same for L-serine as it is for a number of other amino acids. Yet, many of these other amino acids attract certain species of fishes, whereas L-serine, at least to migrating salmon, is a repellent. This suggests that although electrophysiological studies can be useful in estimating the threshold levels of sensory systems to different compounds, they do not indicate how the animal will integrate or react to these stimuli.

Food extracts: Natural food extracts such as the slime excreted from earthworms, and tissue fluid extracts from clams, shrimp, and fish elicit feeding responses in various species of fishes. It is probably the specific combination of amino acids in foods, perhaps with certain other chemicals, that forms a particular "odor profile" recognized by fishes. For example, seven amino acids (glycine, alanine, valine, threonine, glutamic acid, leucine, and serine) were found to be the main constituents of an extract from a marine worm (*Arenicola marina*) that attracted Atlantic cod (*Gadus morhua*) and whiting (*Merluccius bilinearis*). Glycine alone was the most attractive of the amino acids in the mixture. When glycine and alanine were excluded, the mixture was no longer attractive to the fishes. Differences in the attractiveness of certain amino acids to fishes is probably one of the reasons why fish raised on a particular food often resist changing to a new diet.

Because of their relatively poor eyesight, sharks appear to rely heavily on olfaction. They have been shown to respond to the odor of food at concentrations as low as 100 parts per billion, the equivalent to about 250 ounces (about 7.0 kilograms) of food extract in 1,000 railroad tank cars of water. The right- and left-handed turns that result in the typical figure-eight search pattern of sharks are probably the result of the animal moving in the direction of the nostril that receives the greatest stimulus from the odorant. It has been reported that water flowing over a stressed fish elicits a greater response from a shark than water flowing over one that is not stressed.

Body odor and social behavior: A large number of behavioral studies both in the laboratory and in the field have shown that olfaction is very important to fishes in many ways in addition to feeding. For instance, blinded bluntnose minnows (*Pimephales notatus*) have been taught to distinguish between the odors of aquatic plants and invertebrates. When their olfactory epithelium was destroyed, the fish no longer responded to these odors. The ability to discriminate between different body odors has been demonstrated for fishes such as yellow bullhead (*Ictalurus natalis*), which are able to discriminate between members of their own species and between two species of frogs. Also, minnows have been trained to react to body odors of fifteen different species of fishes from eight different families.

Recognition of body odors may be very important in many aspects of the social behavior of fishes. Along with vision and the lateral line system, olfaction may play a part in the schooling behavior of some fishes. A number of studies have shown that olfaction is very important in many phases of sexual behavior and parent–offspring relations. For example, male gobiids (Gobiidae) show courtship behavior when they are exposed to a small amount of water in which gravid females had been placed for only a few minutes. The ovarian fluid was the only body fluid that elicited the courtship response. Male goldfish are also strongly attracted to water from containers holding ovarian fluids and eggs from the same species. The majority of studies concerning the relation between olfaction and sexual behavior in fishes have been done with tropical aquarium fishes because of the ease with which they can be bred in aquaria. There is no doubt, however, that similar relationships exist for common species of fishes.

Crowding factor: A factor or factors that are species-specific and that retard growth, reproduction, and depress heart rate have been extracted from tank water in which fishes have been kept under crowded conditions. This crowding factor has been reported for zebra fish (*Brachydanio rerio*), goldfish, carp, and blue gourami (*Trichogaster trichopterus*). The majority of the heart-rate-inhibiting factor for goldfish and carp was extracted from water with chloroform and was in the neutral lipid fraction. It has been suggested that the factors inhibiting growth, heart rate, and reproduction are either very closely related or the same substance. An interesting aspect of this work that has not been investigated is the significance of the crowding factor in wild populations or in the crowded conditions that often exist in commercial pond culture operations.

Fright substance: The fright reaction—in which fishes stop feeding, concentrate, then seek cover or flee when a member of their species has been injured—was first reported in 1938. Since then it has been reported for many species of ostariophysian fishes and has been associated with olfaction. The fright reaction is initiated by a fright substance, called *Schreckstoff* in German, given off by the injured skin of fishes. The chemical makeup of the compound or compounds is not known; however, histological observations suggest the substance is produced by specialized epidermal cells that do not open to the surface and only release the fright substance when the skin is injured. Fishes killed without injuring the skin do not elicit the response. Also, the fright reaction can be induced by skin extracts of different species. The intensity of the response is generally greater the more closely related the species.

Although the fright reaction is mediated by the olfactory system, fishes with inoperable olfactory systems in the same water or intact fishes in other tanks also display the reaction if they are in visual contact with the fish displaying the initial reaction. Although latency periods of up to five minutes have been reported, dye studies have shown that response to the fright substance generally occurs thirty to sixty seconds after it reaches the fish. Some differences in reaction times may be due to differences in the substance's concentration in the water. Repeated exposure and season also have been reported to affect the degree of response. Fishes that show the fright response can react to very low concentrations of skin extracts. Species of European Cyprinidae reacted to concentrations that were estimated to be in the parts per trillion.

Migration and imprinting: Sockeye salmon (*Oncorhynchus nerka*), coho salmon, chinook salmon (*O. tshawytscha*), and chum salmon (*O. keta*), captured on their spawning runs, were tagged, had their nostrils plugged, and were released with intact conspecifics. The majority of the intact fishes selected their home streams, while those with their nostrils plugged were random in their selection. Although it has been suggested that operations such as plugging the nostrils affect other behavioral traits that may be responsible for the homing behavior, the majority of the evidence suggests that olfaction is very important in the spawning migration of these fishes.

It has been determined that salmon imprint on some odor or combination of odors in their home streams during their fry stage. The imprinting appears to occur when these fishes are undergoing the physiological changes (smolting) that take place prior to their migration to the ocean. Critical periods for imprinting have been reported to take anywhere between thirty-six hours to two months.

A number of questions remain unanswered concerning the role of olfaction in migration. For example, in some streams where fishes make long spawning runs, the chemical stimuli are mixed and diluted by other streams that flow into the home stream. How do these fishes recognize the odor when they return on their spawning run? The answer may be that they imprint on a series of odors as they move downstream instead of just the odor of their natal area. What is the chemical makeup of the odor or odors to which they respond? Although it is known they respond to the organic fraction of the water, the makeup of this fraction is unknown. For species in which there are always some fish in the streams, such as Arctic char (*Salvelinus alpinus*) and Atlantic salmon (*Salmo salar*), it has been suggested that home stream waters are scented by pheromones from nonmigrating members of the population. However, in tests with coho salmon, the odor of juvenile fish was a less powerful attractant than home stream water without the odor of juvenile fish. These and other studies of behavior of migratory fishes suggest there are still a number of unanswered questions concerning the overall imprinting process. They leave no doubt, however, that olfaction is a very important part of this process.

Artificial imprinting: An interesting aspect of the imprinting process has been the use of artificial odorants such as morpholine to imprint salmon in hatcheries. Morpholine (C_4H_9NO) is a colorless, acrid-smelling heterocyclic amine that does not occur naturally in the environment. It can be detected by unconditioned salmon at concentrations of about 1 part per trillion. In Wisconsin, coho salmon were imprinted on this compound for several weeks prior to and after smolting. They were then tagged and released directly into Lake Michigan. When the spawning run began about

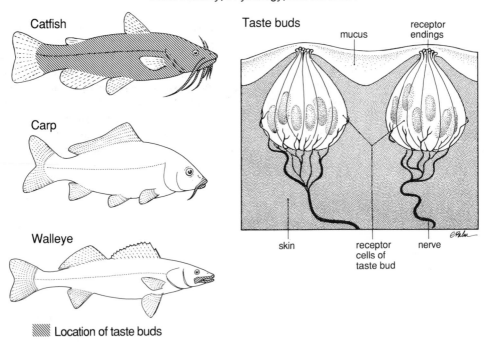

Catfish

Carp

Walleye

▨ Location of taste buds

Taste buds

mucus   receptor endings

skin   receptor cells of taste bud   nerve

Typical vertebrate taste buds occur in fishes, to varying extents and often in widely varying locations.

eighteen months later, morpholine was metered into a stream about 0.5 kilometers (0.3 miles) south of the original release site at a concentration of about 50 parts per trillion. Of the study fish captured in the stream, 218 had been exposed to morpholine and only 28 were control fish. The number of morpholine-treated fish recovered in the stream represented about 2.7 percent of those stocked. This percent return is similar to the percent return for normal homing Great Lakes salmon. Versions of this study were repeated over several years. In each instance, the majority of the morpholine-treated fish returned to the stream scented with morpholine. In one study in which an additional group of fish was treated with phenethyl alcohol (PEA), the majority of the morpholine-treated fish returned to the stream scented with morpholine, and the PEA-treated fish returned to the stream scented with PEA. Brown trout (*Salmo trutta*) and rainbow trout have also been successfully imprinted with morpholine and PEA.

Although the process of imprinting on compounds in water has primarily been studied in salmonids, studies of this process in other fishes could prove to be very rewarding and provide a great deal of information about the relation between olfaction and behavior of fishes. For exam-

ple, in New York a study of walleye (*Stizostedion vitreum vitreum*) in Lake Chautauqua showed that over 90 percent returned to their home stream during their spawning run. Other fishes such as white bass (*Morone chrysops*), striped bass (*Morone saxatilis*), and certain stream-run populations of lake trout (*Salvelinus namaycush*) also return to their home streams. Also, within a lake many species appear to return to the same spawning areas each year. Are these fishes imprinting on certain odor profiles in the water, or are they using other senses to return to their spawning areas?

Morphology and function of taste receptors: In addition to their location in the mouth, taste receptors in fishes are often located in the pharynx, on gill rakers, and on gill arches, and in some fishes they are widely scattered over the outer surface of the body. For example, freshwater fishes such as catfishes have thousands of external taste buds on their bodies, with especially dense concentrations on their barbels. It has been estimated that for a yellow bullhead that is 25 centimeters (about 10 inches) long, there are about 175,000 taste buds on the external body surface and about 20,000 in the mouth and throat. By comparison, there are about 10,000 taste buds on the human tongue. Species such as carp and sturgeons (Acipenseridae) also have large numbers of external taste buds. In some marine fishes, such as the sea robins (Triglidae) and hakes (Gadidae), taste buds are located on specialized pectoral fins that have been modified into feelers. If one considers the poorly lit surroundings in which many of these fishes live, it is reasonable to assume that their ability to literally taste the environment as they move through it is a definite asset in finding food.

Fishes appear to have at least three types of taste receptors: taste buds that are similar to those of higher vertebrates, spindle cells, and free nerve endings. Taste receptors located in the mouth, pharynx, and on gill arches and rakers are similar to the taste buds of higher vertebrates. Dense concentrations of taste buds are also located on the palatal organ, a raised area on the roof of the mouth of cyprinid fishes such as carp, minnows, and suckers. Taste buds consist of three different types of cells: elongate receptor cells arranged somewhat like the segments of an orange, supporting cells, and basal cells. In teleosts, the receptor cells of the taste bud are reported to have one or two microvilli that extend above the surface of the cell. Elasmobranch taste buds have from three to twelve microvilli. Receptor sites for the taste substances are believed to be located on the microvilli, and the mode of action is probably similar to that described for olfaction. Free nerve endings, which are the terminal structures of the taste bud nerve, surround the taste buds and form a plexus in the space between the nuclei of the receptor cells and the basal cells. Specialized epidermal spindle cells, which resemble the receptor cells of the taste bud, are found around the head and body of some minnows. There are many free nerve endings in fishes, and some of these are also believed to act as taste receptors. For example, although the free fin rays of sea robins do not have any specialized taste cells, they do contain a large number of spinal nerve endings. These fishes show positive reactions when the fin rays touch certain chemicals.

The size of specific areas of a fish's brain can often be used as a measure of the importance of a particular sense to the fish. In fishes that rely on taste to find food, the facial and vagal lobes of the medulla are often enlarged. For example, the vagal lobes are larger than the remainder of the brain in suckers, a fish that has a large number of taste buds.

Sensitivity of taste receptors: Gustatory sensitivities of a number of different fishes have been tested for a variety of substances using both electrophysiological- and behavioral-response techniques. These tests suggest that many fishes respond to a wide variety of tastes. For example, taste buds of the palatal organ of carp respond to acetic acid, sucrose, dextrose, levulose, glycine, quinine, NaCl, human saliva, carbon dioxide, and the extract of silkworm papae. For many substances, the sense of taste in fishes is more sensitive than it is in humans. The thresholds for sucrose and salt in minnows are reported to be 2×10^{-5} M and 4×10^{-5} M, respectively, where M stands for molar. These values are about 512 and 184 times lower than those reported for humans. The threshold value for minnows for fructose is about 2,500 times lower than that of humans. Actual threshold values for many substances may be even lower because they are calculated from baselines determined from the fish's response to solutions that theoretically contain zero concentrations of the substances being tested. Sensitive gas chromatographic techniques, however, have shown that in many instances the so-called zero-concentration control substances contain contaminants to which the fishes may respond.

In most fishes, taste is considered a close-range or contact sense used primarily as the final screening process to accept or reject food. However, in some, such as the catfishes, taste is so sensitive that it is also used to detect substances at a distance. As was the case with olfaction, taste receptors in catfishes and other fishes are very sensitive to certain amino acids and combinations of amino acids. Action-potential responses from the maxillary barbel nerve of channel catfish were obtained from concentrations of L-alanine of about 10^{-12} M. Threshold values for glycine, L-arginine, L-serine, L-glutamine, and L-cysteine ranged between 10^{-9} and 10^{-10} M. Behavioral tests with brown bullhead (*Ictalurus nebulosus*) and yellow bullhead suggest that these fishes have a gustatory acuity that is similar to the olfactory acuity of many other fishes. For example, they can use their sense of taste to locate food at a distance of at least twenty-five body lengths. In one study, yellow bullheads rapidly located a food source (liver juice or 0.01 M cysteine hydrochloride) that was dripped into a tank containing 600 liters (about 160 gallons) of water. By means of taste alone the fish exhibited a true gradient searching pattern in the absence of a current. When taste receptors on one side of the body were impaired, the fish would circle to orient the side with intact taste receptors toward the food source. The search patterns suggest these fish were using taste receptors on different parts of their bodies to locate food sources in much the same manner that olfaction is used. Thus, the taste sensation actually begins when these fish are some distance from their food, intensifies as they get closer, and is strongest when the food is ingested.

Respiration

Oxygen is as important to the survival of fishes as it is to the survival of terrestrial organisms. One major difference, however, is that fishes live in an environment that is oxygen-poor relative to terrestrial habitats. Two characteristics of water make it a relatively poor medium for respiration: 1) even under the most ideal conditions, water only contains about 3 percent of the oxygen found in air; 2) water is about 800 times denser than air, so more energy is required to move it across the gills than to move air across the lungs. The fish gill is very efficient at removing oxygen from water. To understand how this efficiency is achieved, it is necessary to examine some of the basic morphological and physiological characteristics of this very impressive organ.

The Respiratory Pump

To get an adequate supply of oxygen, fishes must keep a constant supply of water flowing across their gills. In the relatively primitive lampreys and hagfishes, the gills are basically a series of pouches. With the exception of one species of hagfish, each pouch has an opening that leads from the pharynx to the exterior. Gill filaments and their associated secondary lamellae are attached to the walls of the pouches. In lampreys, contractions of muscles in the outer walls of the pouches pump water out over the gill filaments. When the muscles relax and the pouches expand, water is drawn in. Because adult lamprey use their mouth to attach to prey, it is important that they do not have to use their suckerlike mouths to take in water for respiration. In hagfish, muscle contractions in the velum (part of the foregut), in the gullet, gill pouches, and gill ducts force water through

the external openings of the gill pouches. When these muscles relax, fresh water enters the system through a single nostril on the snout and flows through the nasal passage to the velum.

In most jawed fishes, water is taken in through the mouth during the respiratory cycle. In some sharks, rays, and dogfish, however, there are a pair of reduced gill slits or spiracles located between the first gill slit and the ear capsule. Along with the mouth, these serve as inlets for water during the intake portion of the respiratory cycle. Sharks and rays have separate valvelike openings for each gill slit. In most bony fishes, the gills are covered with an operculum that has a single opening at the rear.

Among teleosts, the volume of the buccal and opercular chambers is increased through a series of muscular actions, and water flows in through the open mouth. Then, through another series of sequential muscular actions, the volume of the two chambers is decreased. By timing this with the closing of the mouth and the closing and opening of the opercular valve, water is pumped over the gill and out through the opercular valve. The pumping action is similar in elasmobranchs, except that some water is taken in through the spiracles, and parabranchial chambers take the place of the opercular chamber.

Certain active fishes, such as tuna, trout, mackerel, and some sharks, pass water over their gills by keeping their mouths and gill chambers open as they swim (ram gill ventilation). This is very efficient, and many species are able to suspend active breathing and rely on ram gill ventilation when they swim at speeds greater than 1.5 kilometers (0.93 miles) per hour. Some fishes, such as adult tuna, mackerel, and the mackerel sharks, have lost the ability to pump water over their gills and suffocate if they are prevented from swimming.

The Gill

Efficient passage of oxygen from water to the blood and passage of carbon dioxide from the blood to water requires that a relatively large surface area of the fish containing a rich blood supply be brought in close contact with the water. The design of the fish gill fulfills these requirements very well. The two main components of the gill that form the sieve through which the water passes are gill filaments and their associated

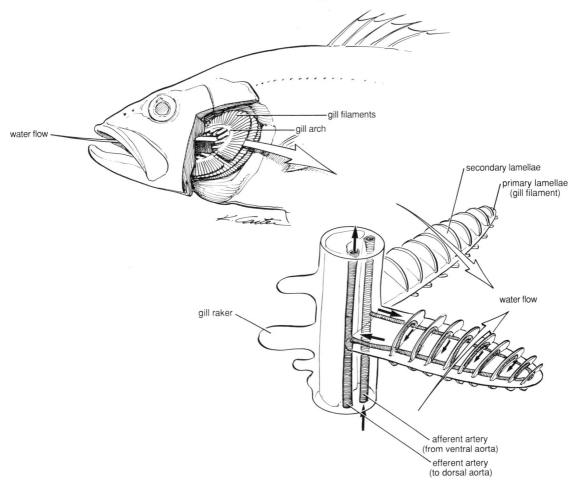

Extreme close-up of one gill arch, showing two gill filaments and their associated secondary lamellae. A unique feature contributing to fish gill efficiency is the countercurrent flow of blood and water.

secondary lamellae. In teleost fishes, two rows of feather-shaped gill filaments extend from each of four gill arches on both sides of the fish. Along both sides of each filament are very small, thin, plate-shaped secondary lamellae. These are the actual sites of respiratory exchange. Each of the secondary lamellae is covered by a thin layer of epithelial cells that are supported internally by pillar cells. Blood in the secondary lamellae flows through capillarylike channels created by the pillar cells. These channels are so narrow that only one red blood cell at a time can literally squeeze through them. In most fishes, the cell wall of the secondary lamellae is only about 1 to 6 microns thick (about 1/68 to 1/11 the thickness of a human hair). Tips of the filaments from one gill arch touch those of the adjacent arch and form the horizontal portion of the sieve. The associated secondary

lamellae form the vertical portion. The channels created by this sieve are about 0.02 to 0.05 millimeters wide and about 0.2 to 1.6 millimeters long. This sievelike design brings almost all the water flowing past the gill in contact with the secondary lamellae and their rich blood supply.

The fact that blood flowing through each gill lamella moves in a direction opposite to that of the water greatly increases the effectiveness of the gill as a respiratory organ. This "countercurrent" exchange system is much more efficient than a co-current system in which blood flows in the same direction as the water. In a co-current system, the diffusion gradient between two fluids is very large when they first come together, but rapidly decreases as the two fluids move in the same direction and approach equilibrium. In such a system, there can never be more than a 50 percent transfer from one fluid to the other. In contrast, by reversing the flow of one of the fluids, there is always a diffusion gradient. Consequently, diffusion from one fluid to the other can exceed 50 percent. The countercurrent system in the gill creates a diffusion gradient that allows oxygen to enter and carbon dioxide to leave the blood the entire time that water is in contact with the secondary lamellae. The value of countercurrent flow was demonstrated in a study in which the flow of water over a fish's gill was reversed. When water and blood flowed in the same direction, the efficiency of oxygen uptake decreased from 51 percent to 9 percent.

The number of gill filaments and secondary lamellae varies considerably among species. Fishes that are more active generally have a greater number of both filaments and lamellae and consequently, a larger total gill area. For example, the total gill area of an active species such as mackerel is about 1,040 square millimeters per gram (45.7 square inches per ounce) of body weight while that of a sluggish fish such as toadfish (Batrachoididae) is about 151 square millimeters per gram (6.6 square inches per ounce) of body weight. The total area of the gill can be ten to sixty times greater than the rest of the surface area of the fish. As impressive as this seems, the total area of the lung of most terrestrial mammals is greater. It would seem to be an advantage for fishes to have an even larger gill surface area. Respiration, however, is not the only function of the gill. It also is an important site of ion and water exchange, and further increases in the surface area of the gill would result in excessive ion and water imbalances that could be fatal to the fish (see the osmoregulation section of this chapter).

Under ideal conditions, fish can extract over 80 percent of the oxygen from water that passes over their gills. The highest figure for the lung in humans is about 25 percent. Unfortunately, the gill also is very efficient at filtering other materials from water, many of which are harmful to fish. The highly solvent nature of water also adds to this problem. Most substances that enter water are dissolved and therefore are subject to being passed over the gill during respiration. Various pesticides, heavy metals, and many other materials can be concentrated in fish tissues through direct uptake from water by the gills (bioconcentration). When these concentrations are added to those accumulated from eating contaminated food (bioaccumulation), it has been shown that fish can accumulate concentrations of these materials that are often thousands, and in some instances millions, of times higher than concentrations in the water. Because the gill is such an efficient all-around filter, it is important for anyone working with fish to maintain good water quality and take every precaution to keep pollutants out of the water.

Fish Blood

Circulation: The bright red color of the fish gill is indicative of the rich blood supply flowing through it. Blood from the heart reaches the gills via the ventral aorta, and then flows to the afferent (inward–flowing) brachial arteries in each gill arch. Branches of the brachial artery form small arterioles that supply blood to the secondary lamellae. As blood flows through the lamellae, individual red blood cells are forced through the narrow channels. At this point, when the red blood cells are closest to the water, oxygen is picked up by hemoglobin in the red blood cells and carbon dioxide is released. After blood passes through the lamellae, it flows via branches of the efferent (outward–flowing) brachial arteries to the dorsal aorta and then to the rest of the body. The blood supplies oxygen to the tissues, picks up carbon dioxide, and then is returned to the gill by the pumping action of the heart to repeat the process.

Hemoglobin and gas exchange: Blood transports oxygen to fish tissues in essentially the same way that it does in terrestrial animals. The principal oxygen–carrying component of the blood is hemoglobin in the red blood cell. Fish blood can carry about fifteen to twenty-five times the amount of oxygen carried in the same volume of water. Of that amount, about 99 percent is carried by hemoglobin. An important characteristic of hemoglobin is that its affinity for oxygen is very sensitive to changes in the concentration of carbon dioxide and pH (hydrogen ions). This affinity decreases as the concentrations of carbon dioxide and hydrogen ions increase. Conversely, as the concentration of oxygen increases, the affinity of hemoglobin for carbon dioxide and hydrogen ions decreases. Because of their effects on the affinity of hemoglobin for oxygen, the concentration of carbon dioxide and hydrogen ions are controlled within very narrow limits by the buffering system of the blood.

Transport of gases from the gills to tissues and

from the tissues to gills is a very complex process that involves a number of intricate biochemical reactions. The brief discussion that follows is only meant as a summary of this process. At the gill, as the red blood cell passes through the secondary lamellae, the number of free oxygen molecules in the water is greater than the number in the red blood cell, so oxygen diffuses into the red blood cell, where it is bound by hemoglobin. Blood leaving the gill is normally about 85 to 95 percent saturated with oxygen, whereas venous blood returning to the gill is about 30 percent to 60 percent saturated. As oxygen is bound by hemoglobin at the gill, the affinity of the hemoglobin for carbon dioxide decreases, and carbon dioxide is released. Because the number of free carbon dioxide molecules in the red blood cell is now greater than the number in the water, carbon dioxide diffuses from the blood through the lamellar membrane into the water.

Blood flows from the gill to the various tissues. In the tissues, the blood enters an environment that contains more free carbon dioxide molecules than the red blood cell due to thé metabolic activities of the tissues. Because of their increased concentration, carbon dioxide molecules diffuse into the red blood cell. The increased concentration of carbon dioxide molecules and their reaction with water also cause an increase in the number of hydrogen ions. The increase in carbon dioxide and hydrogen ions decreases the affinity of hemoglobin for oxygen. This results in an increase in the number of free oxygen molecules in the red blood cell. Because this number is greater than that in the tissues, oxygen diffuses into the tissues. When the red blood cell returns to the gill, the cycle is repeated.

The tendency for carbon dioxide and hydrogen ions to decrease the affinity of hemoglobin for oxygen is called the Bohr effect. This is a very important reaction because it increases the efficiency of oxygen exchange at the tissues. Although it is physiologically advantageous for fish under normal conditions, there are circumstances when the Bohr effect can cause serious problems. For example, anything that causes prolonged vigorous activity, such as capture by hook and line or by a net, can cause a buildup of carbon dioxide and lactic acid in the blood. If these levels increase until they exceed the buffering capacity of the blood, there will be a decrease in the pH of the blood and a corresponding decrease in the affinity of hemoglobin for oxygen at the gill. This can cause fish to die from lack of oxygen and generally occurs when fish are stressed. Active fishes such as trout and salmon have a very strong Bohr effect and therefore are especially sensitive to low pH in their blood. For this reason, when it is necessary to keep fish under crowded conditions, keep their water well aerated.

Strategies for Increasing Oxygen Uptake

Fish can increase oxygen uptake through several short-term responses. One is to increase the pumping or ventilation rate of the gills so more water is passed over the gills. Another is to increase the flow of blood through the gills by increasing the rate or stroke volume of the heart. Efficiency of gas exchange can also be increased by dilating the filamental arterioles, which increases the flow of blood through the secondary lamellae.

These short-term strategies, however, are too demanding energetically to be continued for prolonged periods, and they also can lead to osmotic problems. Fortunately, there are other ways fish can increase their respiratory efficiency for prolonged periods. The most common of these is to increase the number of red blood cells and the concentration of hemoglobin in them. The efficiency with which hemoglobin combines with oxygen at the gill and releases it at the tissues can also be changed. For example, it has been shown that increases in temperature lower the affinity of hemoglobin for oxygen. However, if fish are allowed time to acclimate to the higher temperature, the efficiency of their hemoglobin exceeds that of the nonacclimated fish. The key to long-term adjustment to changes in water quality is that fish need time (generally days or even weeks) to adjust to their changed environment. Most fish normally are not subjected to rapid changes in water quality in their natural environment and therefore cannot make rapid physiological adjustments to such changes. For this reason, rapid changes in water quality of captive fish should be avoided.

Osmoregulation

Osmoregulation involves a number of physiological processes that together maintain the proper internal salt–water balance in fishes. In a manner of speaking, the gill membrane of freshwater and marine fishes is suspended between the blood and the water, media of very different ionic concentrations. To understand the osmotic problems this creates for fish, it is necessary to review the principles of diffusion and osmosis.

Diffusion is the movement of molecules from an area of higher concentration to one of lower concentration. For example, when a bottle of perfume is opened, molecules diffuse from the high concentration in the bottle out into the room. Given enough time, the system will reach equilibrium, with equal numbers of perfume molecules inside and outside the bottle. Osmosis is the diffusion of water through a semipermeable membrane. If a solution of sugar in water is separated from pure water by a membrane permeable only to water, the water will diffuse through the membrane into the sugar solution. The pressure it would take to stop the movement of water into the sugar solution is called the osmotic pressure of the solution. The more concentrated the solution, the greater will be the osmotic pressure. Water will always pass from the solution having the weaker osmotic pressure through a semipermeable membrane to the solution with the higher pressure until the two solutions reach the same pressure. If, as is the case with most living tissues, the membrane is also permeable to certain dissolved substances, water will move from the weaker solution to the stronger one while the dissolved substances will move more slowly in the opposite direction. Again, movement will continue until the two solutions reach the same osmotic pressure.

Because of differences in the salt content of their environments, freshwater and marine fishes face different osmotic problems. In freshwater fishes, the concentration of dissolved salts (ions) in the blood is greater than the concentration in the water (i.e., the fish's body fluids are hypertonic to the water). Therefore, the osmotic pressure of the blood is greater; water diffuses through the gill membrane into the blood, and ions such as sodium (Na^+) and chloride (Cl^-) diffuse into the water. In marine bony fishes, the situation is reversed. The concentration of ions in the seawater is greater than that of the blood (i.e., the fish's body fluids are hypotonic to the water). Water diffuses out of marine bony fishes and ions diffuse in. Elasmobranch fishes maintain an osmotic pressure in their blood that is higher than that of the seawater. Because the osmotic pressure of their blood is greater than the seawater, these fish

behave osmotically like freshwater fishes. This specialized physiological strategy is discussed later in this section. If the osmotic process in each of these groups of fish were not regulated in some way, it would be impossible for them to maintain the salt–water balance of their body fluids at the concentrations necessary for life.

At first glance it would seem that a simple solution to the osmotic problem would be for all fish to have membranes that are impermeable to water and salts. If this were the case, however, the membranes would also be impermeable to respiratory gases and to excretory products such as ammonia. Neither the respiratory nor excretory functions of the gill would be possible. Therefore, as is most often the rule with biological systems, instead of complete optimization of one system at the expense of another, there is a compromise. In this instance, the compromise is that the gill is permeable to water and salts.

In both freshwater and marine fishes, osmoregulation requires the use of energy in the form of adenosine triphosphate (ATP). A large amount of the standard metabolic energy (the energy required for basic metabolic chores such as respiration and digestion) of fish is used for osmoregulation. Consequently, even when a fish is motionless and apparently resting, it still must use considerable energy to maintain its internal salt and water balance. Any factors that increase the osmoregulatory demand above the normal level for more than a short period can be very harmful.

Freshwater Fishes

The kidney: Two basic osmotic problems faced by freshwater fishes are getting rid of excess water and maintaining the proper salt concentration in their body fluids. Excess water is eliminated by an efficient kidney system that produces a very copious, dilute urine. The kidneys of most fish are reddish, paired longitudinal structures found just below the vertebral column. The basic structure of the kidney is the nephron, which consists of a renal corpuscle and a convoluted tubule. The renal corpuscle consists of an open-ended, double-walled capsule (Bowman's capsule) and a mass of capillaries called the glomerulus, which are tightly coiled inside the capsule. The tubule from each nephron drains into the main collecting duct for that kidney and

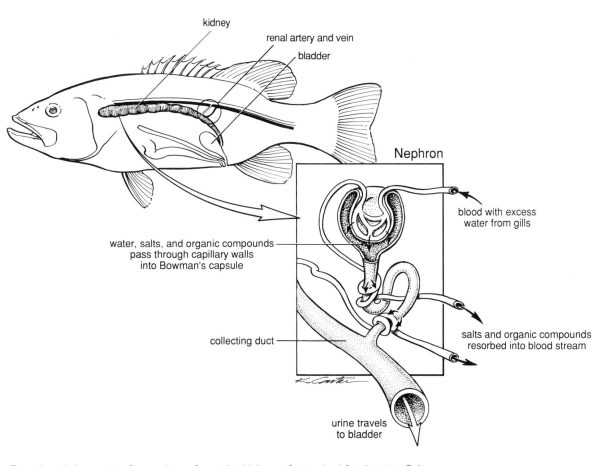

kidney

renal artery and vein

bladder

Nephron

blood with excess
water from gills

water, salts, and organic compounds
pass through capillary walls
into Bowman's capsule

salts and organic compounds
resorbed into blood stream

collecting duct

urine travels
to bladder

Functional elements of a nephron from the kidney of a typical freshwater fish.

then to the exterior.

The most important function of the kidney of freshwater fishes is to eliminate excess water. One might think of a freshwater fish as a leaky boat, with the major leak at the gills, and the kidneys as the pumping system used to control the leak. Relatively speaking, the leak is quite extensive. Terrestrial vertebrates only produce up to 1.5 percent of their body weight per day as urine. Freshwater bony fishes, however, may produce more than 20 percent. Along with water, ions and organic materials such as glucose and amino acids are also filtered out of the blood as it flows through the glomerulus. However, as the filtrate passes through the convoluted tubule of the kidney, most of the ions and organic materials are reabsorbed back into the blood, while the majority of the water is passed out of the fish as a very dilute urine. For example, almost all of the Na^+ in the filtrate is reabsorbed; however, usually less than 50 percent and sometimes as little as 5 percent of the water passes back into the blood.

The gill: Although the reabsorption of ions in the kidney is very efficient and produces a very dilute urine, there is still a substantial loss of ions because so much urine is produced. To counter this loss and the loss of ions through diffusion at the gill, ions are absorbed from the water by the gills. Principal sites of this uptake appear to be the membranes of the secondary lamellae and specialized "chloride cells" located on the gill filaments, primarily at the base of the secondary lamellae. Although there still are many unanswered questions concerning the details of this very complex process, the most prevalent theories suggest that uptake of Na^+ and Cl^- ions against a concentration gradient is mediated by a class of enzymes called ATP-ases. These enzymes are located within the cell membrane and are often referred to as "ion pumps." They require energy in the form of ATP for the active transport of ions through the cell membrane against the concentration gradient. They also require the presence of counter-ions. For example, the inward movement of Na^+ seems to be coupled with the outward movement of hydrogen ions (H^+) and ammonium ions (NH_4^+), and the inward movement of Cl^- is apparently coupled with the outward movement of the bicarbonate ion, HCO_3^-. These reactions are essential not only in the osmoregulatory process but also are an important part of the acid-base and excretory function of the gill, because they serve as a mechanism for the elimination of excess H^+, HCO_3^-, and NH_4^+ ions.

The transepithelial potential, an electrochemical gradient across the gill membrane, is another aspect of gill physiology that plays an important part in osmoregulation. The charge of the electrical gradient is such that the inside of the gills of freshwater fishes is generally negative relative to the outside. It has been suggested that this difference in charge between the inside and the outside of the gill membrane decreases the tendency for positively charged ions to diffuse out through the gill membrane. It also decreases the amount of energy needed for the active transport of positively charged ions from water into the blood.

Marine Fishes

Because the blood of marine bony fishes has a lower osmotic pressure than the seawater (i.e., is hypotonic), their osmotic problem is the reverse of that of freshwater fishes; water diffuses out through their gills and salts diffuse in. Solving the water part of their osmotic problem is relatively simple. Marine bony fishes replace water lost through their gills by drinking 7 to 35 percent of their body weight in seawater per day. As the salinity of the water increases, they compensate by drinking more seawater. Unfortunately, although this replenishes the lost water, it also increases the salt concentration of their blood. The salt content of the blood is increased further by salts that diffuse in through the gills. Because the osmotic problems of freshwater and marine fishes differ, the roles of the kidney and gills as osmoregulators also differ. In marine bony fishes, the kidney functions to conserve water and excrete salts, and the gills also excrete salts.

The kidney: Although the basic structure of the kidney of marine bony fishes is similar to that of freshwater fishes, it is generally less developed. The glomeruli, the principal sites where water is filtered from the blood, are generally smaller and fewer in number than those in freshwater fishes. In fact, a number of marine fishes have no glomeruli. Also, the distal segment of the kidney tubule is missing. The distal segment is the principal site for the reabsorption of monovalent ions such as Na^+ and Cl^- in freshwater fishes. Because their glomeruli are generally less developed and their kidney tubules are more permeable to water, marine bony fishes produce a urine flow that is only about 1/10 to 1/20 that of freshwater fishes.

While the principal function of the freshwater bony fish kidney is to get rid of water, it appears the main function of the marine bony fish kidney is to excrete magnesium sulfate. The majority of the divalent magnesium and sulfate ions taken in during the drinking process are excreted in the feces. However, about 20 percent of these ions do enter the blood. Available evidence suggests that almost all the magnesium and sulfate ions that are absorbed into the blood are excreted by the kidney. Although other ions such as sodium and chloride, are found in the urine, it is the gill that serves as their major excretory pathway.

The gill: The principal osmotic function of the gills of marine bony fishes is active transport of excess Na^+ and Cl^- from the blood to the water. Chloride cells appear to be the major site for transport, and ATP-ases appear to play a major part in the sequence of biochemical events involved in the actual transport process. There are, however, major differences in the transport process between freshwater and marine bony fishes. For one, in marine bony fishes, the ATP-ases are polarized in the opposite direction (they transport salts outward), and the counter-ion for the ATP-ase responsible for the transport of Na^+ appears to be potassium instead of the H^+ or NH_4^+ ions. For another, the transepithelial potential of marine bony fishes is such that the inside of the gill membrane is positive relative to the outside. Energetically this may make it easier for these fish to excrete positively charged ions against the inward diffusion of ions from the seawater. For fishes such as salmon that migrate between fresh and salt water, the direction of ion transport at the gills reverses when they move from one environment to the other. Changes in the osmoregulatory functions of the kidney in such fishes are more a matter of quantity than direction. For example, urine output of salmon that move from salt to fresh water will generally increase from 8- to 12-fold as a result of the increased influx of water and the resulting increase in the glomerular filtrate.

Elasmobranch Fishes — A Special Strategy

Elasmobranchs have evolved an osmotic strategy that is somewhat different than that of either freshwater or marine bony fishes. Basically they maintain a very high concentration of urea and other nitrogenous waste products in their blood. This is made possible, at least in part, because the gills of these fish are relatively impermeable to urea. Also, most of the urea and other nitrogenous waste products in the glomerular filtrate are reabsorbed from the kidney tubules. The result is that the urea concentration in the blood of most elasmobranchs is about 2 percent to 2.5 percent, while that of most other vertebrates is only about 0.01 percent to 0.03 percent. In most animals, a blood urea concentration above 0.5 percent is lethal. The combination of salts at concentrations similar to those in bony fishes and nitrogenous waste products increases the osmotic concentration of the blood so that it is slightly hypertonic to the seawater. As a result, these fish have a glomerular filtrate and a urine flow that is higher than that of marine bony fishes. Energetically this is probably a more efficient osmoregulatory strategy than that of marine bony fishes because the osmotic gradient between seawater and the blood is less for elasmobranchs than for marine bony fishes.

The Stress Response

Although most people intuitively understand what stress and the stress response mean, both of these terms have been given a number of definitions. For this discussion, we will consider a stress or stressor to be the stimulus acting on a fish and the stress response to be the physiological response of the fish to the stimulus. Defining the point at which a stimulus acting on a fish actually becomes a stress is difficult because, even under the best of conditions, fish are subjected to a wide variety of chemical and physiological changes. At what point does any one or any combination of these changes become severe enough to become a stress? Although there really is no accurate way of defining this point, we will consider a stimulus to be a stress if it requires a physiological response from the fish to adapt to the stimulus.

Some familiar examples of the alarm phase of the stress response are the sudden jolt of energy you feel when you realize you have almost fallen asleep while driving, the superhuman feats of strength performed by people in emergency situations, and the "pumped up" feeling an athlete gets before the game. These responses to stress situations are part of a series of physiological reactions called the general adaptation syndrome. This syndrome is divided into three phases: the alarm reaction, when stress hormones are released; a stage of resistance, during which adaptation occurs; and, if the animal cannot adapt, a stage of exhaustion followed by death. These reactions were initially observed in terrestrial vertebrates. However, similar, but not identical, reactions have been observed in fish.

The Stress Hormones

Organs of stress hormone production: The stress response is mediated by two groups of hormones: catecholamines, which are associated with the more immediate reactions to stress; and corticosteroids, which are more related to the adaptive phase. The catecholamines, adrenalin (epinephrine) and noradrenalin (norepinephrine), are produced by chromaffin tissue and nervous tissue. Chromaffin tissue, which is analogous to the medullary (inner) portion of the adrenal gland of higher vertebrates, is located in the anterior dorsal part of most fishes, embedded in the head kidney. Nerve tissues associated with the production of catecholamines are the nerve endings (adrenergic nerve endings) that innervate organs such as gill, and the heart. Catecholamines produced by chromaffin tissue are carried by the blood to various target organs. Catecholamines produced by adrenergic nerve endings affect the organs innervated by the nerves and apparently have little effect on concentrations in the blood. Corticosteroids (mainly cortisol) are produced by interrenal tissue, which in most fish is also found in the area of the head kidney. Interrenal tissue is analogous to the cortical (outer) portion of the adrenal gland.

Method of release: Release of the stress hormones is controlled in two ways: by direct nervous stimulation; and by releaser substances and adrenocorticotrophic hormone (ACTH) produced by the brain and the pituitary gland, respectively. Catecholamines are released by direct stimulation of nerve tissue. Therefore, the effect of these hormones on organs innervated by adrenergic nerve endings is very rapid and generally represents the first phase of the stress response. Release of catecholamines from chromaffin tissue is also controlled by the nervous system. Increased levels of these hormones can be detected in the blood of fish within several minutes of the initial stress. Release of the corticosteroids is controlled by a releaser substance produced by the hypothalamus in the brain and by ACTH produced by the pituitary gland, which is located directly beneath the brain. When the presence of a stress is sensed by the brain, a releaser substance produced by the hypothalamus stimulates the release of ACTH by the pituitary. The ACTH is carried by the blood to the interrenal tissue, where it stimulates the secretion of corticosteroids, which are carried by the blood to the various organs they affect.

Principal effects: Some of the most obvious effects of catecholamines are on the circulatory system—increased cardiac output and constriction or dilation of blood vessels. In some instances, epinephrine causes dilation of a particular set of blood vessels and norepinephrine causes constriction; in others, both hormones cause the same effect. The effect of these hormones on heart rate is not completely understood. In some studies, stress or the injection of catecholamines caused a decrease in heart rate, while in others it caused an increase. The most consistent response reported has been an increase in stroke volume, which results in an overall increase in blood pressure. During periods of stress, one of the main effects of catecholamines is to increase the flow of blood to specific organs. For example, at rest only about 65 percent of the secondary lamellae on gill filaments are perfused by blood. However, because of the increased stroke volume of the heart and the catecholamine-caused dilation of filamental arterioles, about 95 percent of the secondary lamellae can be perfused by blood in a stress situation. This results in a greater efficiency of oxygen uptake. Therefore, more oxygen is available for the increased metabolic demands associated with the stress.

Catecholamines also appear to affect osmoregulation and metabolism. There is good evidence that increases in the concentration of catecholamines in blood affect osmoregulation by increasing the permeability of the gill to water and ions. In freshwater fishes, this results in an increase in the influx of water and an increase in the loss of ions from the body fluids to the water. For marine bony fishes, it means a loss of water and an increase in the influx of ions. The major metabolic effect associated with elevated levels of catecholamines in the blood is the breakdown of glycogen to glucose in the muscle and the liver. The breakdown of glycogen in muscle occurs very rapidly and provides an almost instant energy source. This, together with the increased oxygen supply, provides the fish with more energy.

The principal effects of the corticosteroids appear to be on osmoregulation and metabolism. In teleosts, these hormones appear to affect osmoregulatory processes acting on the gills, kidney, and intestinal tract. For example, increases in the concentration of cortisol in blood appear to affect active transport of ions at the gill by facilitating uptake of certain ions in freshwater fishes and excretion of ions in marine teleost fishes. The most prominent effects on metabolism involve breakdown of glycogen to glucose and the breakdown of proteins. Amino acids that result from the breakdown of protein are used to form new proteins or glucose. An example of this type of metabolic effect occurs during the nonfeeding migratory life stages of salmonids. During migration, high levels of cortisol facilitate the breakdown of protein to glucose (gluconeogenesis), which serves as a major energy source for these fishes.

Acute and Chronic Stress

The initial response to an acute stress is called the "fight or flight" response. When a fish is pursued by a predator, the stress response results in an almost instantaneous increase in the available energy, which may allow it to escape. Other energy-requiring functions such as osmoregulation are temporarily shut down so as much energy as possible can be channeled into getting away from the predator. In some ways, the purpose of the stress response is similar to that of the supercharger in a car. It is a short-term measure to produce large amounts of energy to deal with emergency situations. Most acute stresses are short-term. The fish either escapes or is caught. If it does escape, the stress is relieved, and the fish can recover from the adverse osmotic effects. In the wild, fish can alleviate stresses caused by changes in water quality by moving to a more favorable habitat.

In captivity, however, fishes are often subjected to relatively long periods of stress. The physical act of handling, changes in water quality, and crowding are all stress factors associated with captivity that have been shown to elicit the stress response. The trauma or "fright" associated with being pursued and captured and then held in a foreign environment can result in fishes being stressed for prolonged periods. A common cause of death in stressed captive fish is osmotic imbalance (osmotic shock). This occurs as a result of the increased blood flow through the secondary lamellae and the increased permeability of the lamellae to water and ions. A fish can correct the osmotic problems caused by a short-term stress, but if the stress is prolonged, death often occurs from several hours to several days after the initial stress. Stress also suppresses aspects of the immune response, such as phagocytoses and intracellular killing by macrophage cells. A frequent result is the outbreak of a disease from several days to about two weeks after fish have been moved from one environment to another.

Studies with salmon suggest fish can become conditioned to the stresses associated with capture. In one study, blood cortisol levels in coho and chinook salmon that had been netted and released twice daily for periods of ten, fifteen, and twenty days were lower than those of unhandled controls. In our work, we have found that when wild bluegill are acclimated to capture, there is little mortality associated with moving them. It has been suggested that hatchery fish generally have lower resting levels of stress hormones than wild fish. Although this makes them easier to handle, they may be less alert and therefore at a disadvantage when they are released and forced to compete with wild fish.

A fish exposed to a chronic stress can either compensate and adapt to the stress, or fail to compensate and die. Even if a fish does compensate, its performance capacity will be reduced during the period of compensation. In many instances, even after compensation, the perfor-mance capacity will be lower under the new set of conditions. For example, fish can adapt to rather wide changes in water temperature, but within this broad temperature range there is a preferred range in which the fish generally grows and performs best. The further the temperature is above or below this range, the poorer will be the overall performance. Changes in the allocation of energy to the various systems within the fish are one of the principal reasons for the changes in performance. The systems responsible for the basic metabolic functions necessary for life, such as the nervous system, the respiratory system, and the systems that control osmoregulation, have a higher priority for energy than growth or reproduction. Under conditions of little or no stress, there is an energy surplus that can be put into growth and reproduction. However, the further conditions are from the preferred range, the greater the stress and therefore the greater the amounts of energy the fish must use to make the proper physiological adjustments. Thus, the energetic cost needed by the fish to maintain itself under the new set of conditions may be such that growth and reproduction are adversely affected. Examples of the adverse effects of chronic stress on captive fishes are common and are often caused by poor water quality. If the water quality or other factors causing the problem are corrected, growth generally improves, and in many instances the fish begin to reproduce.

Alleviation of the Stress Response in Captive Fish

There are several ways to decrease the stress response in captive fishes. One is to decrease the overall "awareness" of the fish. The less aware a fish is of external conditions and factors such as discomfort and pain, the lower will be the stress response. One way to accomplish this is with anesthetics such as Tricain Methane Sulfonate (MS-222). This technique is widely used when fishes are moved in hatchery operations. Elimination of a fish's visual awareness of the surroundings also appears to have a calming effect. For example, cortisol levels in stressed juvenile steelhead trout began to decrease within seven hours in fish held in the dark, but remained high in those held in the light.

Adding salts such as sodium chloride to the water in which freshwater fishes are transported and cooling the water are techniques widely used in hatcheries to increase survival. Salt reduces the effects of the stress response by lowering the osmotic gradient between the water and the blood. This, in turn, reduces the influx of water and the loss of salts and therefore reduces osmotic shock. This technique has proven to be very successful in diminishing the stress response for both warm- and cold-water fishes. We have found that the addition of about 1 percent sodium chloride to water used to transport bluegill from local ponds to our laboratory increased survival from less than 50 percent to over 90 percent. Cooling the water that fishes are transported in slows their metabolic rate, which tends to reduce the stress response. Also, in most instances the stress response will be lower in fishes captured from colder water. Consequently, the best time to capture most fishes in the wild is in the spring, fall, or winter when the water is cool.

Selected References

Ali, M. A., ed. 1979. *Vision in Fishes*. New York: Plenum Press.

Bond, C. E. 1979. *Biology of Fishes*. Philadelphia: Saunders College Publishing.

Brown, M. E. 1957. *Fish Physiology*. Vol. 1. *Metabolism*. Vol. 2. *Behavior*. New York: Academic Press.

Health, A. G. 1987. *Water Pollution and Fish Physiology.* Boca Raton, Fla.: CRC Press.

Hoar, W. S., and Randell, D. J., eds. 1969. *Fish Physiology.* Vol. 1. *Excretion, Ionic Regulation, and Metabolism.* Vol. 2. *The Endocrine System.* Vol. 3. *Reproduction and Growth, Bioluminescence, Pigments, and Poisons.* 1970. Vol. 4. *The Nervous System, Circulation, and Respiration.*1971. Vol. 5. *Sensory Systems and Electric Organs.* Vol. 6. *Environmental Relations and Behavior.*1978. Vol. 7. *Locomotion.* 1979. Vol. 8. Energetics and Growth. 1983. Vol. 9. *Reproduction.* Part A. *Endocrine Tissues and Hormones.* Part B. *Behavior and Fertility Control.* 1984. Vol. 10. *Gills.* Part A. *Anatomy, Gas Transfer, and Acid-Base Regulation.* Part B. *Ion and Water Transfer.* New York: Academic Press.

Kleerekoper, H. 1969. *Olfaction in Fishes*. Bloomington: University of Indiana Press.

Love, M.S., and Cailliet, G. M., eds. 1979. *Readings in Ichthyology*. Santa Monica, Calif.: Goodyear Publishing.

Marshall, N. B. 1966. *The Life of Fishes.* Cleveland: World Publishing Co.

Northcutt, R. G., and Davis, R. E., eds. 1983. *Fish Neurobiology*. Vol. 1. *Brain Stem and Sense Organs.* Vol. 2. *Higher Brain Areas and Functions*. Ann Arbor: University of Michigan Press.

Pickering, A. D., ed. 1982. *Stress and Fish*. New York: Academic Press.

Smith, L. S. 1982. *Introduction to Fish Physiology.* Neptune City, N.J.: T.F.H. Publications.

Tavolga, W. N., Popper, A. N., and Fay, R. R., eds. 1981. *Hearing and Sound Communication in Fishes.* New York: Springer-Verlag.

Toshiaki, J. K., ed. 1982. *Chemoreception in Fishes*. New York: Elsevier Scientific.

Nutrition and Feeding of Tropical Fish

Robert A. Winfree

Fish are an extremely diverse group. It is therefore not surprising that this diversity is reflected in their food habits. No single food will meet the needs of all species of ornamental tropical fish at every stage of their life cycle. The more common aquarium fish have been selected for fish-keeping, at least in part, for their flexible food habits. Yet many beautiful species kept only rarely in aquaria require special foods that are available in the wild but which often are not easily provided in aquaria. A basic understanding of the food habits of fish in the wild can aid in the selection of foods appropriate for a wide variety of fish in aquaria.

Few species of fish can be neatly categorized as strict carnivores (meat eaters) or herbivores (plant eaters). Furthermore, the type of food selected in the wild may vary seasonally with its availability. Most species kept in aquaria are more or less omnivorous, seeming to require, or at least to prefer, a variety of animal and vegetable matter.

The blue tilapia (*Tilapia aurea*) is an example of a species with highly flexible food habits. When they are young, these fish commonly feed on zooplankton, but may also prey upon smaller fish. As tilapia grow larger, they usually filter food from the water or grub through bottom sediments for food. Still, all sizes of tilapia readily accept a wide variety of prepared foods. Flexible food habits

such as these are a real asset when several species are reared in the same aquarium.

Many highly specific terms, not used in this chapter, have been coined to describe particular food habits. However, to apply such terminology broadly can give false impressions about the food habits of a given species. Few species of fish have precisely the same food habits as any other species, and few retain the same food preferences as they age. Feeding habits often are reflected in morphology as well. A more generalized classification, based largely on the method of feeding rather than on the items consumed, follows.

Fish that nibble at plants or that pick at small plankton or benthic animals can be called grazers. Many common aquarium species fit in this group. The young of most species feed on zooplankton. Generalized grazers such as guppies and mollies feed on a wide variety of plant and animal foods. Other species can be very selective. For example, the blue tang (*Acanthurus coeruleus*) and some loricariid catfish (such as *Farlowella*) eat almost nothing except for algae in the wild. Adults of some marine angelfish and butterfly fish graze primarily on sponges and coral polyps, respectively. The bluehead wrasse (*Thalassoma bifasciatum*) and several other marine species commonly remove

Many fish are morphologically adapted for particular modes of feeding. A knowledge of their food habits in nature will enable the aquarist to select appropriate foods for captive fish. Top: the suckermouth catfish (*Plecostomus*) is a vegetarian scavenger; middle: the tiretrack eel is a predator; bottom: a scat (*Scatophagus*) feeds on both plant and animal matter.

and consume crustacean parasites from other species of fish, but these "cleaner fish" also feed on zooplankton or benthic organisms.

Fish that concentrate planktonic plants or animals by straining the water are called filter feeders. Finely spaced gill rakers enable them to separate food from the water efficiently. Some species use brushlike pharyngeal teeth to further concentrate or select foods. Many African cichlids are filter feeders, as is the paddlefish (*Polydon spatula*).

Bottom feeders or scavengers include some carps, loaches, sturgeon, and catfish, which are able to obtain nourishment from plant and animal debris (detritus) and from the invertebrate animals that live in the sediment. Bottom feeders often have sensitive fleshy lips and ventrally positioned sucking mouthparts. Barbels are a type of sensory appendage common to bottom feeders. Some species even have taste buds in their abdominal skin. The type of sediment influences feeding success. Soft sand and mud are easily processed, but coarse bottom gravels can interfere with feeding. Sediment makes up 10 to 20 percent of the stomach contents of some bottom-feeding fish. The sediment may carry nutrients in the form of fine particles or surface films and may also aid in the digestion of algae or other foods.

Only a small percentage of fish are totally predatory, but some of these are adapted for preying on a specific type of organism. A consistent source of small fish, worms, or live plankton may be required to keep such species well fed. However, some predatory species can also be trained to accept frozen foods or even dry diets.

Several species, from more than a dozen freshwater or marine families, have evolved into highly specialized parasites which feed on the fins, scales, or body fluids of other fish. Some species have even evolved to mimic the appearance and behavior of harmless fish. Such behavior allows a parasite to approach its host with less likelihood of being recognized as a threat.

Feeding Stimuli

Most species of fish can eventually be trained to eat prepared diets, but not all diets are equally acceptable to all fish. Even the most nutritious diet cannot maintain a fish that fails to recognize the food. Recognition is affected by instinct and by training. The interaction of feeding stimuli is complex. Hunger, security, and state of health are important motivational factors that are easily overlooked. Temperature, water quality, and illumination (duration, direction, and intensity of light) are especially important. Cyclical rhythms—including seasonal, reproductive, tidal, and solar cycles—control feeding activity in the wild and may persist in tank-reared fish.

Many fish will not feed after being moved to a new tank, whereas others will approach food within minutes of capture from the wild. A new fish that fails to feed for a day or two is not cause for concern, but longer fasts may require changing the diet or the conditions under which the fish is held.

If the tank is arranged to mimic a natural environment, the fish may be calmed and therefore may feed better. For some, this can mean as little as including a piece of plastic pipe or a clay flower pot for shade, but providing a well-landscaped tank for timid fish is never a mistake. Aquaria can be so brightly lit that some fish refuse to feed. The intensity of illumination should be reduced if fish stay in hiding or cluster in the corners.

Fish locate food by various visual, chemical, or mechanical clues. After locating a potential food item, fish frequently examine and even taste it before deciding to swallow. Many predatory species swallow prey quickly and regurgitate unsuitable foods later. Because of differences among species, broad generalizations about the most important sensory characteristics are inappropriate. Fish can often be trained to accept unique foods. Recognition of these new foods may not involve previously important clues.

Flavors: Olfaction and taste are important for most species and may be especially significant for bottom feeders and others in turbid water. Although there are specific anatomical receptors for taste and smell, flavors must dissolve in the water to be detected. Fish do not necessarily respond to the same chemical stimulants as humans. Although some food ingredients stimulate fish to feed, a number of others are actually repellent. Certain amino acids, nucleotides, and carboxylic acids have attractant properties. Sweets and fats are less effective. Mixtures of synthetic amino acids approximating the composition of natural foods are more powerful as feeding stimulants than are single compounds. Combinations of organic chemicals may soon be developed to augment or extend the flavor of natural ingredients in prepared foods. However, synthetic chemicals alone do not yet elicit activity equivalent to that of fresh foods. Seafoods are especially strong attractants. Experimentation re-

Many predacious fish are not easily trained to accept nonliving food. A consistent supply of live foods may be required to keep such fish healthy. The large South American tiger shovelnose catfish (*Pseudoplatystoma fasciatum*) is among these species.

mains the best method to determine acceptable flavors for fish foods.

Sound: Sound is important in the feeding behavior of some fish, such as channel catfish (*Ictalurus punctatus*). Scattering food over a small part of a catfish pond can cause a frenzied search for food by fish throughout the pond. Fish commonly congregate in the feeding area even before the fish farmer arrives, possibly because they have detected the sound of his footsteps or his truck.

Buoyancy and color: Buoyancy is also important for some species. Fish adapted to feeding at the surface may not pursue food which has settled to the bottom. Some bottom dwellers may not surface to take floating food either, but the majority of aquarium fish are less picky. Shallow-water species have good color vision, and color may be an important feeding stimulus. Lighter-colored foods seem to be more acceptable during initial training. Some species seem to prefer red or green foods.

Feeding of predatory fish: Many predacious fish are not easily trained to accept nonliving foods. Predators commonly utilize scent or touch to locate food or are triggered to feed by visual stimuli. For some species, the peculiar characteristics of live food are especially important to trigger feeding. Detection of prey can involve visual clues, vibrations, and even changes in electrical fields.

The food preferences of a predator newly introduced to the aquarium can be inferred by offering a variety of potential live foods. A method for training a predator to accept prepared foods is as follows. Provide a preferred live food until the fish is eating regularly, then gradually reduce the amount offered. Substitute freshly killed food organisms on subsequent days. Some manner of simulating movement, such as dropping the food into water currents at the tank surface, may be necessary to attract the attention of the fish. Introduce soft meaty foods for a portion of the killed organisms in later feedings. If prepared foods are still refused, try coating foods with a puree of organisms to which the fish is already accustomed. Training is more easily accomplished with small individuals before they have developed inflexible habits. Sometimes the training process can be accelerated by introducing other individuals of the same or similar species that are already trained to take nonliving foods.

There are undoubtedly many unique fish with such highly specialized food habits that they are impractical for the hobbyist to keep. For example, some marine fish feed on sponges, coral polyps, marine algae, or live zooplankton. It can tax an aquarist's ingenuity to supply the authentic article or to find substitutes. Fish recently collected from the wild can be especially difficult to train. As technology advances, appropriate prepared foods will no doubt be developed for species now thought to require living food.

Functional Anatomy of Feeding and Digestion

The digestive system of a fish is relatively simple when compared to that of higher vertebrates such as humans. However, there is great variation between species. Mouthparts of fish frequently are specially modified for unique food habits. Some species, such as seahorses and mormyrids that feed on zooplankton or small bottom invertebrates, have tubular sucking mouthparts. Such fish may be incapable of eating larger foods. In contrast, some predators capture and swallow prey much larger than would at first seem possible. Special joints enable them to open their mouths to enormous proportions while their esophagus expands to allow large prey to be swallowed whole. Some predators, such as the sea-trout (*Cynoscion nebulosus*) and the sargassum fish (*Histrio histrio*), cannot distinguish their own kind from other prey. Hatchery production of such cannibalistic species is particularly difficult.

Not all species have teeth, but when they are present they are often adapted for special functions. For example, predators such as gar and pike have needlelike teeth designed to incapacitate and restrain prey. On the other hand, predatory catfish generally have small teeth. These catfish, like many other predators, do not bite small prey but rather inhale them with suction created by the mouth and gill flaps. The teeth of piranha are used to slash or cut flesh from prey rapidly, whereas the pacu, a closely related group, have teeth adapted for crushing nuts and fruits. The front teeth of parrotfish are fused into a powerful clipper, used for feeding on coral and coarse algae. Parrotfish also have a set of pharyngeal teeth at the rear of the mouth cavity to grind the food before swallowing. The pharyngeal teeth of tilapia are almost like a brush and help to separate minute foods from debris or water.

Anatomists define the stomach as the part of the gut that secretes acid. Hydrochloric acid secreted from stomach glands activates enzymes that digest protein. Tilapia are able to digest algae because stomach acids rupture the algal cells and release nutrients. The stomach of predators is often sac-shaped and can accommodate enormous amounts of food. Another stomach adaptation is found in puffer or blowfish, which can rapidly inflate their stomach with air or water. This adaptation protects the puffer against predation by increasing its size. A heavily muscularized portion of the stomach functions like a gizzard in some herbivorous and detritivorous species, including *Prochilodus, Citharinus,* mullet, and some surgeonfish. Digestion may be impaired in *Labeo* and other cyprinid fish, as they have no true stomach.

Partially digested food passes from the stomach into the intestine, where it is digested further and nutrients are absorbed into the body. Bile salts produced by the liver and stored by the gallbladder neutralize stomach acids and emulsify dietary fat. The pancreas secretes enzymes that digest carbohydrates into the intestine. The intestine of herbivorous species is often elongated and is more complicated than that of carnivorous fish. The inner surface of the intestine may be folded or have fingerlike projections that extend into the intestinal cavity. Such structures increase surface area for absorption. Small sacs, called caecae, open into the intestine of many species and also serve digestive or absorptive functions. Foods not completely digested during passage through the intestine leave the body through the anal opening.

Food Nutrients and Their Functions

Protein and energy: Protein is the major nutrient required for growth of fish and, on a dry-weight basis, makes up most of the body structure. The importance of proper protein nutrition cannot be overemphasized. The essential components of proteins are amino acids, which are used by fish to synthesize new body tissues and enzymes. In addition, proteins are a significant source of dietary energy. Proteins vary in their ability to support growth, depending upon their source and processing. Nutritionists often use purified casein (milk protein) or a blend of casein and gelatin to provide the protein in research diets. Although limited amounts of casein or gelatin are sometimes included in formulated feeds, seafoods and fish meals provide the best combination of digestible amino acids. Feed dried at very high temperatures loses nutritional value, because essential amino acids bind to other components and become unavailable.

Fish are especially efficient at converting food to body tissues, so they need less food to grow than do many other animals. However, since the diet of fish contains relatively little carbohydrate matter, the amount of protein is high as a percentage of the diet. Because protein is the most expensive part of the diet, it is important to feed just the right amount and type for best growth. Excessive dietary protein is ultimately excreted as ammonia by the fish.

The amount of protein required in the diet depends upon several variables, including the species of fish, the growth rate, and the amount of natural food available. In a typical production scheme for trout, five or more foods are fed in sequence as the fish grow. These dry foods vary in particle size and level of protein to match the specific requirements of each size of fish. A similar phase-feeding regimen has also been developed for catfish. Tropical fish hatcheries are not usually so large as to make phase feeding necessary for cost control. However, by selecting ap-

Newborn fish such as these channel catfish (*Ictalurus punctatus*) grow rapidly and require a very rich food for maximum growth and survival.

propriate foods for each stage of production, producers and hobbyists can benefit from accelerated growth rates and improved fish health and survival.

Fish fry and larvae grow rapidly and require a very rich diet for maximum growth and survival. In the wild, protein makes up 50 percent or more of their diets on a dry-weight basis. High-protein foods are important in the hatchery, too. The protein requirement of fish gradually decreases as they gain in size. Young adults of many species can be reared on foods containing 35 to 40 percent protein. Less protein may be appropriate for some older fish but not usually for breeders. Foods produced for pond use usually contain only 25 to 35 percent protein, because pond fish are expected to forage for much of their food. Protein requirements are dramatically affected by water temperature. When fish are held at cool temperatures, their growth rate falls, so lower dietary protein levels may be appropriate.

Fats and essential fatty acids: Food components that can be separated in the laboratory with solvents such as ether or chloroform but not with water are called lipids or crude-fat. Several different classes of chemicals can be isolated from lipids, including triglycerides (true fat), fatty acids, steroids (precursors to hormones), phospholipids, and several important vitamins and pigments. We commonly refer to crude-fat or lipid simply as fat for convenience.

Fats can supply energy for normal body needs, sparing proteins for growth. Carbohydrates can also serve this function. If a feed is well designed, it will supply just enough energy for maximum growth without producing fatty fish. Tank-reared fish are especially prone to fattiness because they expend little energy searching for food. Thus, the ratio of dietary fat to other food nutrients is important. Most production feeds contain only 5 to 8 percent fat on a dry-matter basis. Even so, greater amounts of fat are appropriate for very young fish, for carnivorous species offered foods low in carbohydrates, and for egg development. Additional processing steps are sometimes used to raise the fat content of hatchery diets to 12 percent or more.

Nutritionists caution against using too much saturated fat, such as beef tallow, in fish feeds. Hard fats are not digested easily and can interfere with metabolism at temperatures below their melting point. Dry commercial feeds are usually balanced for level and type of fat, but the amount of fats in fresh meats can be deceiving. Beef heart can vary from about 15 percent to more than 50 percent on a dry-matter basis. Only the leanest cuts of meat should be used as fish food, and these should be carefully trimmed to remove all visible fat, or a nutritional imbalance is likely to result. Reproductive success of some fish species is reduced when nonessential saturated animal fats or vegetable oils make up too much of the diet.

Several carnivorous species require a source of fish oil in their diet to supply essential polyunsaturated fatty acids of the linolenic group. Recent evidence indicates that specific dietary fatty acids may also be important to reproductive success (survival of fertilized eggs). Fats commonly make up 20 to 50 percent of the moisture-free fraction of fish eggs. Commercial fish breeders and hobbyists have reported for many years that certain foods are especially valuable for conditioning fish to breed. However, the specific nutritional basis of fish reproduction has remained largely undetermined. Diets containing polyunsaturated fats derived from marine fish or invertebrates (krill, brine shrimp, clams, squid, or annelid worms) are increasingly being recommended for conditioning freshwater and marine fin-fish and shrimp to breed. The specific lipid composition of the fish and invertebrates used as fish food varies with the site of harvest, probably reflecting local differences in algae and their other foods. Lipid composition is at least one factor responsible for differences in quality among strains of brine

shrimp (*Artemia salina*) used as food for larval fish.

Feed mills may use fish meals that include fish oils, or they may add oil to the diet. Several suitable oils, including cod-liver oil, are available to feed manufacturers. All the fish oils tend to turn rancid through oxidation, and they eventually can become toxic. Rancid fish oil has an especially disagreeable odor, although it can be hard to distinguish from that of the fresh oil. Feed manufacturers add chemical preservatives to the oil as protection.

Lecithin is a substance commonly separated from soybean oil. It contains phospholipids that are valuable nutrients and it facilitates the dispersion of dietary fat during digestion. A specific requirement for a lecithin supplement has not been proven for fishes, but it is sometimes added to foods at a level of 1 to 2 percent for the reasons mentioned above.

Carotenoid pigments: Fat-soluble carotenoid pigments (carotenes and xanthophylls) are responsible for the yellow, orange, red, and green colors of the skin, flesh, or eggs of many fish and crustaceans. The brilliant blue and violet colors of crustaceans also result from carotenoids, although the blue colors of fish have other origins.

There is mounting evidence that some of these pigments are important nutrients. Survival of fertilized eggs to hatching has been correlated to the concentration of pigments in the embryonic yolk and in the prespawning diet of the female fish. Carotenoid pigments may serve a protective function for delicate membranes and other sensitive tissues in the developing embryo. Astaxanthin, a red carotenoid pigment, is also reported to have reproductive functions in male fish. The amount of vitamin E (and other antioxidants) needed in the diet is higher when oil supplements or carotenoid pigments are included.

Some foods that are rich in pigments include brine shrimp, krill, and other species of zooplankton. Bright-red, orange, or yellow roe from fish and crustaceans is sometimes sold in fish markets. Roe is also rich in essential oils. Commercial sources for xanthophyll pigments include red fish oil, meals or extracts of fish roe and crustaceans (shrimp, krill, or crawfish), marigold petals, algae, alfalfa, corn gluten, annatto, paprika, and others. Purified or synthetic carotenoid pigments are also available. These are not the same as the food colors commonly sold in food markets.

Analytical determination of the carotenoid pigment content of an ingredient will not necessarily indicate the value of the ingredient in fish food. The various carotenoid pigments have different physiological properties that can even change during processing and storage. Furthermore, species of fish differ in their ability to metabolize specific pigments. Controlled feeding trials, using several species of fish, remain the best method to compare ingredients for color enhancement.

Carbohydrates: Carbohydrates such as starch and sugars make up 20 to 40 percent of most commercial foods. They apparently are not essential for growth, but they are inexpensive sources of energy. In fact, carbohydrates are so commonly distributed in

A catfish with ocular cataracts. Cataracts can cause blindness in fish given foods containing inappropriate levels of certain minerals, vitamins, or amino acids.

feedstuffs that it would be difficult to exclude them completely from a practical food. Most fish tolerate 30 to 40 percent carbohydrate in their diet, but a condition similar to diabetes results when unbalanced foods are fed. Trout are normally given foods containing less than 20 percent digestible carbohydrate. Too much carbohydrate in the diet of very young fish can prevent them from obtaining enough of other essential nutrients.

Raw starch, the principal nutrient contained in cereal grains, is digested incompletely by fish. High levels of raw starch in the diet can even interfere with the digestion or assimilation of other nutrients. However, starch that is gelatinized by cooking is more digestible and often is used as an economical binder to stabilize foods against disintegration in water. Floating foods usually contain high levels of carbohydrate to facilitate processing.

Vitamins: Vitamins are organic compounds that serve as catalysts for many biochemical reactions in body tissues. Deficiency of almost any vitamin can result in retarded growth and increased susceptibility to disease. Some vitamins may be present in adequate amounts in common feedstuffs, but the cost of supplementing vitamins in a prepared food is low compared to the consequences of a vitamin deficiency. Current practice is to supplement the food with a prepackaged blend

of vitamins, called a premix, when fish are reared in confinement or in high-density pond cultures. Purified research diets are commonly supplemented with fourteen vitamins: A, D_3, E, K, thiamine (B_1), riboflavin (B_2), niacin, pyridoxine (B_6), pantothenic acid, folacin, B_{12}, ascorbic acid (C), biotin, and inositol. Choline (a non-vitamin micronutrient) is also added. Ingredients that are naturally rich in conjugated B-vitamins, such as yeast and whey, are also commonly included in hatchery diets (even though newly developed vitamin premixes may make these unnecessary for most applications). This is because the conjugated forms of these vitamins tend to be more resistant to the leaching effect of water than are manufactured supplements.

Changes in culture practices or diet can change vitamin requirements. Fish stocked at low densities in ponds may obtain enough vitamins from the wild plants and animals that supplement their diet. This is one reason why simple foods that are unsuitable for aquarium use may be adequate for use in ponds. The requirements for several B-vitamins increase when fish are treated with antibiotics, because microbes that can synthesize vitamins in the gut are killed by antibiotic treatments. The requirement for vitamin A may increase when fish are exposed to stress. Certain dietary ingredients (including fish or vegetable oils, raw seafoods, and raw egg white) also in-

The mouthparts of the sheepshead (*Archosargus probatocephalus*) are functionally adapted for scraping and crushing. In this species and others, indigestible shell or sand in the diet may be important for the digestion of coarse foods such as algae. *Photo by:* Tom Smoyer, Harbor Branch Oceanographic Institute.

crease requirements for specific vitamins (including vitamins E and K, thiamine, and biotin). Vitamin E is another lipid component that has been correlated to hatching success. Choline chloride can react to reduce the potency of water-soluble or fat-soluble vitamins in a vitamin premix. Consequently, it is commonly added to manufactured foods separately from the vitamins. Several mineral nutrients can also catalyze vitamin breakdown; therefore, mineral supplements normally are not combined with the vitamins in a premix.

Feed mills generally oversupplement vitamins to compensate for losses anticipated during processing, storage, and feeding. Nevertheless, certain vitamins are among the most perishable components in the food. Vitamin C, for example, can be depleted from a dry food within a few months after manufacture. The feed industry is working to improve the shelf life of foods by developing stabilized forms of vitamins and by modifying processing and packaging methods. However, the best protection against vitamin deficiencies is probably to vary the diet regularly and to buy foods in small quantities that can be used up within a couple of months. The vitamin content will be prolonged by storing extra food and vitamin supplements in a freezer. Feeding fresh or frozen vegetables and live foods can serve to supplement the diet of aquarium fish.

Minerals: Minerals are required in the body for bones, teeth, scales, and tissue fluids. They also serve a variety of supporting functions in body chemistry. For example, iron in the hemoglobin molecule enables blood to carry oxygen to the cells. Calcium and phosphorus are the major minerals most likely to be lacking in fish diets. Fish extract some calcium from hard waters, but fish kept in soft water need calcium in their food. Natural forms of phosphorus found in plants are not available to fish in unplanted aquaria. Bone from fish or meat meal is an excellent source of both calcium and phosphorus, as are several manufactured supplements. Several other minerals, collectively called the minor minerals (because they are required in minute amounts), may also need to be supplemented in the food. Among these are manganese, iodine, copper, zinc, iron, cobalt, selenium, and possibly chromium. The minerals magnesium, sodium, potassium, and chloride are also essential in the diet, but usually are present in adequate amounts. Many minerals are poisonous if present in excess, so mineral supplements should not be added to the food indiscriminately.

It was mentioned earlier that silt and sand commonly make up 10 to 20 percent of the diet of bottom-feeding fish. These are probably ingested inadvertently with food, but nonetheless sand is important to certain fish. Species with a functional gizzard or pharyngeal mill can use grit to abrade algae and other foods for digestion. A hard surface on which to gnaw may also be important for marine parrotfish, triggerfish, and porgies, which commonly graze on coral rock or barnacles. Such fish may fail to prosper in an aquarium environment when denied their regular foods. A novel solution to this problem, incorporating food in a plaster matrix, is sometimes used in public aquaria. (For further information, see under "Plaster blocks.")

Fiber: Most research into fish nutrition has been

Sample formulas for tropical fish foods[1]						
	Gel	Pellet	Pellet	Paste	Paste	Granule
Protein content (%)[2]	40	45	35	45	35	50
Ingredients (%)						
Meat, fish, or shrimp	42	50	50	72	72	—
Fish meal	—	27	14	10	—	80
Rolled oats	11	18	31	9	20	—
Whole-wheat flour	—	—	—	—	—	11.5
Frozen spinach	10	—	—	—	—	—
Unflavored gelatin	5	—	—	—	—	—
Cod-liver oil	2.5	—	—	2	1	3.0
Bone meal	2	—	—	2.5	2.5	—
Vitamin mixture[3]	2.5	2.5	2.5	2.5	2.5	2.5
Guar or cellulose gum[4]	—	2.5	2.5	2.0	2.0	2.5
Sodium propionate	—	—	—	—	—	0.5
Vitamin E (IU)[5]	150	250	250	150	150	400
Water	25	—	—	—	—	—

[1] Directions for preparation are included in the text.

[2] The protein content given is for a food reduced to 10 percent moisture (air dry). Protein content is listed so that these formulas may be compared to dry commercial foods. Variations in ingredients can cause the finished foods to vary from the protein content shown.

[3] The amount of vitamins required depends on the composition of the particular supplement to be used. Follow the manufacturer's recommendations. Review the information presented under "Shelf life and vitamin refortification" in the text.

[4] Added as a preservative to inhibit growth of mold during drying.

[5] Added as an antioxidant to inhibit fat spoilage.

conducted with predatory fish, especially trout and catfish. These species have little, if any, ability to digest ingredients that are high in fiber content. Alfalfa meal, for example, is almost completely indigestible by channel catfish. When low levels of dietary fiber were added to the diet of fish fed purified research diets, growth rate and food conversion efficiency improved. However, higher levels of fiber reduced feeding vigor and growth rate. Dietary fiber facilitates the passage of food through the gut of higher animals, and it may serve such a function for fish as well. Young fish are less able to tolerate foods of low nutrient density and high fiber content, apparently because they are not able to obtain adequate amounts of nutrients.

Unidentified nutritional factors: Throughout the history of fish culture, there have always been certain food ingredients which apparently contained nutrients for which no requirement was yet proven, but which seemed to improve growth or reproductive success when added to a balanced diet, even in relatively small amounts. The active components of many such feedstuffs already have been identified. Consequently, nutritionists are able to select ingredients on the basis of known nutritional content rather than be bound by inflexible traditional-formula feeds. Premixed supplements can be added to guarantee that a feed formula will meet known requirements for vitamins, minerals, and fatty acids. Development of a suitable vitamin premix has made it possible to replace ingredients such as fermentation products (whey, yeast, and distillery by-products), wheat germ, and liver, previously essential for

Vitamins are essential for normal growth. The food offered to the four fish on the right had been stored for too long before use. The larger fish on the left was fed the same food after it had been refortified with vitamin C.

feeding trout. Suitable feeds have also been designed without fish meal using improved methods for processing soybeans and for synthe- sizing amino acids, along with new information about fatty acid requirements. Recent discoveries about the significance of fish oils and carotenoid pigments in reproduction have already been discussed.

Not all nutritional factors have yet been identified, however. Sometimes particular fish will refuse their food and only a change of diet will entice them to feed again. For example, some fish, including scats, surgeonfish, marine angelfish, and some loricariids, that feed heavily on algae or other plants, languish in captivity if vegetable matter is deleted from their diet. Leafy vegetables, such as spinach or lettuce, are suitable for some fish but not for all.

Biochemists recently have documented pharmacologic properties in literally thousands of extracts taken from marine plants and invertebrates. Many of these compounds have never been isolated and studied before. It is tempting to speculate that the special nutritional needs of some fish could stem from a requirement for previously unknown nutrients or from the protective action of certain foods against disease. A better understanding of the nature of unidentified factors should result from more complete knowledge of nutritional biochemistry and of the differing requirements among species. However, for the foreseeable future, prudent fish culturists and

feed manufacturers will continue to include certain ingredients in the diet as sources of unidentified nutritional factors. Specific recommendations for ingredients are discussed throughout this chapter.

Choices in Fish Foods

The development of scientifically prepared foods was spurred by a combination of economic factors and scientific accomplishments early in the twentieth century. The foods used in fish hatcheries of that time were made up mostly of fresh meat by-products, especially liver. Raw fishery products and dried feedstuff such as cereal grains and oilseed meals were not utilized effectively until the scientific bases of nutrition and the composition of potential feedstuffs were better understood.

It is now possible to prepare completely nutritious foods from a variety of common and inexpensive ingredients. Vitamins and other nutrients contained in fresh or live foods can be incorporated into prepared foods as well. Live foods and fresh meats or seafood are no longer essential for rearing most fish, although they remain of use for some species.

The natural foods of fish are highly perishable. The actions of degradative enzymes, microbes, and atmospheric oxygen rapidly change food composition unless it is stabilized by preservation. Fresh foods can lose nutrients and become unpalatable or even poisonous within a few hours of harvest, so preservation is essential unless a continuous supply of fresh food is available. Freezing and dehydration are the most common methods of preservation today, but moist shelf-stable foods are under development. Quick deep-freezing remains the best method of preservation, retaining most of the original nutritional content. However, frozen foods can spoil if allowed to thaw, and open packages must be carefully sealed to prevent freezer burn (uncontrolled dehydration and resultant rancidity).

Dried foods have the advantages of greater convenience and lower cost. Several of these have been developed, including dried natural foods such as plankton and several prepared foods. Most large-scale commercial fish farms use dried foods of one kind or another. Dried natural foods are not the equal of their live or quick-frozen counterparts. Essential nutrients, especially vitamins, can be lost during preparation. Essential fatty acids can turn rancid after drying if they are not stabilized by antioxidants. Such dried "treats" can still be palatable to fish, and they are popular with aquarists for this reason. However, they should not be used exclusively. Feeding a variety of dry prepared foods, perhaps supplemented with fresh meats and live plant and animal foods, is a practical solution (for further information, see under "Making Your Own Fish Foods" and "Live Foods and Their Culture"). Freeze-dried foods are superior to those dried by any other method and are generally worth their higher cost. However, these are also subject to oxidative rancidity if not stabilized by antioxidants, as are all dry foods that contain any unsaturated fats.

Prepared foods (flakes, meals, pellets, and granules) are made up of a variety of ingredients that are ground into a fine powder and sometimes processed further. Fine grinding blends minute amounts of essential ingredients, such as vitamins, evenly throughout the food. The simplest dry foods are ground meals. Meals are comparatively inexpensive to produce and are widely used for feeding small fish in ponds. A simple formula used on some Florida tropical fish farms combines one or two parts of oats and wheat by-products to one part of fish meal. Even after fine grinding, the ingredients in these foods can separate during feeding. Simple meal foods generally are not suitable for extended use in confined tanks and aquaria.

Pelleted foods are prepared by subjecting a meal to moisture, heat, and pressure to bind the ingredients together. With special processing and careful selection of ingredients, pellets can be made to resist disintegration in water for several hours. Several additives developed for the food processing industry can be used to bind ingredients that normally do not stick together. Among the most effective binders are gelatinized starch, alginate, and guar. Lignin, cellulose gums, and a number of other binders can be used where extended stability is not required. Water-stable foods are especially valuable for fish that eat slowly. Different formulas and processing techniques are used to produce floating or sinking pellets. Pelleted foods can be crumbled to make smaller granules, which then can be separated into several sizes for feeding to different-sized fish. Frequently, however, different formulas are prepared for fish of different sizes.

Flakes can be prepared solely from the same ingredients used in pelleted foods or fresh foods can be included. The ingredients are combined with water and blended into a thick slurry. The mixture is pressed onto a hot rotating drum, where it dries quickly and is flaked off by a knife blade.

Selecting Prepared Fish Foods

With the myriad of fish foods available, it may seem difficult to determine which are appropriate for your fish. Cost should not be the major consideration when selecting fish foods. Inappropriate or inferior foods do not save money in the long run.

Most popular species of aquarium fish will eat flake foods. Because flakes soften quickly without disintegrating in the water and do not sink rapidly, they are especially suitable for aquarium use. As the flakes gradually pass through the water column, they are progressively fed upon by top-feeding species (such as guppies), mid-water species (such as tetras), and bottom-dwelling species (many catfish). Granules are also fine for many fish. However, granules have the disadvantage of sinking rapidly and becoming lost in tanks with gravel bottoms. The excess food will spoil unless cleaned up by catfish, loaches, or other scavengers. Sinking pellets are a good choice for some bottom-feeding fish, because the pellets are relatively dense and permit a large fish to eat

Foods made at home can be varied to suit many difficult species, but must be carefully formulated to be well balanced.

hatched fish. They are also suitable for the young of many other species, but are too rich for most adult fish. Larger pellets can be crushed for feeding to adults of smaller species.

Good dry foods contain fish meal, grain, vitamins, minerals, and preservatives. Fishery products, such as fish or crustacean meals, are prominent in the list of ingredients. Soybean meal and meat and poultry by-product meals can make up part of the formula, but should not totally replace fish meal. Vitamins, minerals, binders, and preservatives are included to ensure complete nutritional content or to protect nutrients against deterioration during storage. Many of the substantial improvements in the quality of recent fish foods result directly from the use of these additives. Special ingredients sometimes are added as a source of unidentified nutritional factors. These include meats (liver and glandular meals), seafoods (fish roe, liver, and solubles), fermentation by-products (yeast, whey, and brewer's or distiller's products), wheat germ, algae, and others.

Fish may be attracted by colored foods, and therefore some manufacturers color foods which are naturally tan or brown. Only a few ingredients, however, contain the specific carotenoid pigments that can intensify skin colors of fish. Crustacean meals (shrimp, crab, lobster, brine shrimp, copepods, and krill), salmon skin or roe, red oil extracts of fish or crustaceans, marigold petals, alfalfa, paprika, certain algae, and the concentrated pigments astaxanthin and canthaxanthin are among these.

A few commercial foods contain androgenic steroid hormones, such as testosterone, which enhance skin color by accelerating development of breeding colors. Testosterone accelerates the growth rate of some fish, but can interfere with reproduction. Steroids added to the diet will sometimes result in reproductive sterility or can even reverse genetically female fish to functional males or vice versa. The specific effect of these steroid hormones depends on several factors including, but not necessarily limited to, the specific

its fill more easily than with flakes. Floating pellets are good for cichlids, catfish, goldfish, and other large fish that feed at the surface. Floating pellets are especially useful in garden pools or ponds, because the fish can be watched as they surface to eat, reducing the likelihood of overfeeding. Floating pellets do not disintegrate as quickly in the water as do most sinking pellets.

Foods for fish reared in aquaria, tanks, or cages must be formulated to supply all their nutritional requirements. These foods are called "complete." Some are referred to as "hatchery foods" or "cage foods." They tend to be more costly than incomplete foods intended for pond use, because they contain higher quality ingredients and are supplemented with extra vitamins. Foods recommended for feeding to pond fish such as minnows or catfish are not, as a rule, appropriate for tank use. Serious deformities related to nutritional deficiencies can result when they are fed to fish in aquaria, but more often the fish just waste away or succumb to disease.

Larger farm-supply stores frequently carry economical feeds that are suitable for pond culture. Commercial minnow foods are appropriate for small live-bearers. Catfish foods are accepted by many larger fish, including exotic catfish, cichlids, and goldfish. Trout foods are made in as many as ten sizes. Several different formulas are available as granules of different sizes. The smallest granules are designed for feeding to newly

When raw seafoods are used as feed regularly, fish can develop thiamine deficiency, resulting in failure to feed, clamped fins, and nervous incoordination. These are signals to change the diet. Here, a snook shows wasting symptomatic of hypothiaminosis.

chemical used, dose, duration of treatment, and age at start of treatment. Steroids can also be absorbed from the water by fish.

Even among complete foods there can be significant differences, so it is best to experiment and to alternate foods regularly. Fresh or frozen foods make a welcome change for most fish and are essential for a few species.

Shelf life and vitamin refortification: Sunlight and heat accelerate nutrient breakdown in dry foods, as does moisture. To prolong the potency of nutrients, supplies should not be stored in the culture room but rather in an air-conditioned room or in a freezer (below 0 degrees C or 32 degrees F). Dry fish foods that have accidentally become wetted should be discarded immediately. Bacteria and fungus can grow rapidly on the nutrients in wet fish food, creating risk of food poisoning.

Producers of bulk feeds for fish farming recommend using a product within two or three months of manufacture. The same time frame is probably appropriate for aquarium foods once the container has been opened. The main reason for this is because certain vitamins and oils in the feed gradually break down on contact with oxygen, a change that cannot be detected easily. Manufacturers can extend the shelf life of their dry fish foods through special processing and packaging. One technique is to top-dress the food with vita-

mins, fats, and preservatives just before packaging. Another is to seal the food in airtight containers with a controlled atmosphere such as nitrogen to exclude air. Of course, bulk commercial foods repackaged for sale under a new label are less likely to receive the same care.

When dry foods are sealed in airtight containers and stored in a freezer, they will remain fresh for much longer than those stored at room temperature. Refrigerators are less suitable for storage of dry foods because the food can become damp from condensation. Dry food stored for too long can still be used if it is alternated with fresh supplies. It is also possible to resupplement the food with vitamins, although it is not always practical to do so. Vitamin C is the least stable ingredient in complete foods. Refortification of this vitamin alone can improve growth rate and survival of fish fed an otherwise complete food.

To refortify small quantities of food, vitamins are blended into the food with a liquid carrier (20 to 50 milliliters of carrier to every kilogram of food [0.32 to 0.80 fluid ounces per pound]). The carriers most often used are oils, especially cod-liver oil. Tallow (beef fat) and soybean oil work well in many cases. Aquarists should review the information on fats presented earlier in this chapter. Liquified gelatin is also used as a carrier, although vitamins dissolved in water tend to be less stable than those mixed into oils.

Bulk vitamins can be obtained by mail from pharmaceutical or chemical companies. Vitamin pills, bought at a pharmacy and reduced to a fine powder with a mortar and pestle, are equally suitable as fish food additives, only less convenient. The purity of manufactured vitamins is equal to, or better than, natural supplements available from "health food" outlets. For 1 kilogram (2.2 pounds) of dry food, 50 to 500 milligrams of vitamin C should be used. The higher level provides a safe margin for vitamin loss in the water. Although fish may receive some benefit from multivitamin supplements designed for others (1 to 2 doses per kilogram or 2.2 pounds of food), the

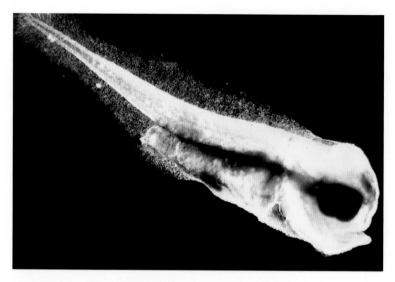

The young of many oviparous species, including *Scatophagus argus,* are so small that dried foods or *Artemia* nauplii are too large as a first diet. Smaller live foods such as rotifers, nematodes, mollusk larvae, ciliates, or even phytoplankton are essential to rear such species.

vitamin requirements of fish are not really comparable to those of humans or other animals. Vitamin supplements designed specifically for fish have recently appeared on the market.

To prepare for resupplementing a dry food, oil carriers should be warmed, if not already liquid. Gelatin must be dissolved in water for use (2.5 grams of unflavored gelatin in 50 milliliters of boiling water, or 0.1 ounce to 1.8 fluid ounces). The gelatin solution or oil should be cooled to 60 degrees C (140 degrees F) before adding vitamins. The vitamins should be blended into the carrier. Then the mixture should be blended evenly throughout the food. A kitchen mixer works well for small quantities. A clean cement mixer is suitable for larger batches. To prevent spoilage, extra vitamins, fish oils, and refortified foods should be frozen in airtight freezer bags until used. These same procedures can be used to incorporate measured doses of medicine into a dry food. Medicated foods are especially useful for treating sick fish in ponds.

Making Your Own Fish Foods

Aquarists rearing large numbers of fish may prefer to prepare some of their own foods. Foods made at home can be varied to suit many difficult spe-

cies, but are not always well balanced. A good practice to follow is to vary the formula occasionally and also to feed complete, prepared foods. Several adaptable formulas are outlined in the accompanying table. Moist gels, pastes, and pellets can be prepared in a typical kitchen. Dry foods are more difficult to prepare properly. The hobbyist should take advantage of the excellent selection of commercial foods available rather than try to make all of his or her own food.

Gelatin diets: Among the simplest and most acceptable of home-prepared foods are gels. Formulas vary widely and are generally imprecise. A typical formula that uses gelatin as the binder is shown in the table. To prepare this food, mince raw meats or seafoods in a meat grinder or blender with a little water to facilitate blending. Blend in the gelatin, oats, and bonemeal and simmer the mixture in a saucepan until just about to boil (about 90 degrees C or 194 degrees F). Do not add vitamins until the mixture has cooled to 60 degrees C (140 degrees F). If the temperature is too high, vitamin potency will be reduced. If too low, the gel may not set properly. Dry fish foods can be added before the gel sets if a little more water is used in the formula. Minced vegetables can also be used if the amount of water is reduced. Pour the mixture into trays or freezer bags, and refrigerate it until hard, then store it in the freezer. Freezer bags with zipperlike closures are especially convenient; they can be filled to a depth of about 2 centimeters (about 1 inch) and laid on a flat surface to cool. Such bags are easily stacked to maximize freezer space. Gel foods can be chopped or shredded after they have set. If the formula is modified, the amount of binder should be limited to only as much as is needed to make a firm gel (usually 1 to 5 percent by weight). Excessive amounts of gelatin in the diet are not good for fish, because although it is pure protein, it is also poorly balanced. When a gel fails to harden, the cause is likely to be either insufficient cooking or too much handling during cooling.

Paste foods and moist pellets: An early method for preparing traditional paste foods was to combine minced meat, oatmeal, and cool water in a saucepan and to bring this mixture to boiling while stirring constantly. After the mixture cooled to 60 degrees C (140 degrees F), other dry ingredients or vitamins were added and the food was refrigerated or frozen until feeding. The oatmeal and other cereals used as binders in early formulas required cooking, but heat is not essential now that there are binders capable of setting in cold water. Such food binders facilitate processing and can delay disintegration of the food in water, but they may also impair

Clear glass or plastic bottles are usually chosen for small working cultures of microalgae, ciliates, rotifers, and *Artemia*.

the flavor and digestibility of the food if used to excess. With a little experimentation, paste foods suitable for many species can be prepared in meat grinders, food processors, or blenders. Review the information provided in the table and elsewhere in this chapter for selection of ingredients. Even with binders, paste foods tend to dissipate and to foul the water more quickly than other types of food, so they should be used carefully in aquaria. Among the better choices of binders are alginates, pregelatinized starches, guar, and cellulose gum. These are used at a 1 to 5 percent level in the food, except for starch, which may make up 10 percent or more of the diet. Instant potato flakes and instant cereals for infants can be used as feed binders, but they require some heat to achieve maximum binding. Pastes can be frozen in thin layers and chopped into small pieces for later feeding.

Moist pellets are especially suitable for larger fish. The formulas shown in the table are prepared by passing the ingredients through an electric meat grinder several times. Most grinders come with an assortment of chopper plates having holes of varying diameter through which the food is extruded. Several sizes of pellets can be made by using different plates.

Plaster blocks: Dry foods are sometimes embedded in a plaster of paris matrix for feeding parrotfish and other coral grazers. A common application of this for the home aquarium is the

so-called "weekend feeder" block. A typical formula calls for about 250 milliliters (8 fluid ounces) of high-purity dental plaster (100 percent calcined gypsum or calcium sulfate hemihydrate), 125 milliliters (4 fluid ounces) of tap water, and 125 milliliters of dry feed. The plaster and water are blended until smooth (like cookie dough), and the dry feed is folded in. Expanded (floating) feed is recommended, because compressed (sinking) pellets can expand in water and cause the plaster to crumble. The soft mixture is pressed into a plastic ice cube tray or a lightly coated (with cooking oil) muffin pan. After hardening (from fifteen to sixty minutes), the blocks can be removed from the mold and stored in a refrigerator. Feed blocks are wet and therefore perishable. Plaster can increase water hardness in aquaria and thus may affect pH.

Dry starter foods: A formula for a granulated dry starter food (the first food given to fish) is given in the table, since these foods often are not available in the small quantities needed by aquarists. The dry ingredients used in a starter food should be ground to a fine powder. A food blender will work. Use a fine kitchen strainer to separate large particles for regrinding. Blend water into the dry mixture with a food mixer, about 250 to 400 milliliters per kilogram of dry food (4 to 6 fluid ounces per pound). Then spread the moist mixture in a thin layer (about 3 millimeters or 1/4 inch deep or less) to dry at room temperature in

An inexpensive plastic box with a thin layer of appropriate medium works well for culturing terrestrial species such as microworms and white worms.

front of a fan. If an oven is used to dry fish food, the nutrient content of the food is likely to be severely reduced. Redistribute the food from time to time so that it will dry evenly. Dry food can be crushed with a rolling pin and sieved through screen strainers for particles of appropriate size. Sodium propionate (a digestible food preservative) is included in the formula to inhibit decay while the food is drying.

Ingredients: Fresh fish are a source of high-quality protein and are widely accepted among fish. The best source is frozen fillets sold for human consumption. Roe (fish eggs) is a good supplement, although it tends to fall apart unless it is cooked or is combined with a binder. This tendency may be an advantage when very small fish are fed. Fatty species of fish, such as carp, herring, mullet, and mackerel, should be avoided for use in feedstuffs, as should spoiled fish and offal. Because fish fillets are deficient in vitamins and minerals, they should not be fed as the only food item. Caution should be employed with raw fish, as they frequently contain an enzyme that can destroy the vitamin thiamine, and because they can be contaminated by disease organisms. Feed mills precook fresh fish to reduce these risks.

Some fish will readily accept squid, clams, and scallops when they seem to ignore everything else. The precautions for shellfish are the same as for raw fish. Fresh shrimp or krill are very well accepted by fish. The shell should be removed to feed shrimp to small fish, or the whole body can be ground into prepared foods. Whole shrimp or krill can be fed to larger fish.

Meat is a valuable addition to the diet, but only lean cuts are acceptable and these must be trimmed of all visible fat. Aquarists like to use beef heart because it resists deterioration in water and is easily chopped to size. Beef heart can vary from about 15 percent to more than 50 percent fat on a dry-matter basis, depending on how carefully it is trimmed. Liver, spleen, or other pulpy meats should be blanched in boiling water to coagulate fluids that could otherwise foul the tank water. Poultry may be acceptable to fish that refuse the strong flavor of beef heart or liver.

Hard-boiled egg yolk, which is easily broken into fine particles, often is fed to fry as a first food. If whole eggs (fresh or dried) are included in a food, they should be precooked. Many dairy products dissipate quickly in tank water and spoil, making them unsuitable for direct aquarium use. However, whey, casein, skim milk, cheese rind, and other by-products can be included in compounded diets for flavor and nutrients.

Live aquatic plants will meet the special needs of herbivorous fishes, but the most desirable plant species often are not available. Vegetables are an acceptable substitute for many fishes, are commonly available, and are generally of good quality. They can be offered raw, although many species prefer softer blanched or prefrozen vegetables. Bright-green or yellow varieties are among the most nutritious. Blanched spinach and crumbled frozen lettuce are common favorites.

Quality control: There is more to food manufacture than picking the right ingredients to blend together. Changes that occur during processing and storage can destroy certain nutrients. Excessive heat during processing can inactivate vitamins (especially vitamin C) and can reduce protein quality by inactivating certain amino acids (especially lysine).

Fish and vegetable oils may spoil unless excluded from oxygen by careful packaging and protected by antioxidant additives. Certain mineral nutrients and some foods, including blood meal, can catalyze and accelerate spoilage by oxidative rancidity. Antioxidants protect dietary fat from rancid spoilage. Ethoxyquin is commonly used (at a 0.0125 percent level) to stabilize dry commercial foods. Vitamin E can be substituted in homemade foods, at a level of 400 to 500 International Units (IU) per kilogram (2.2 pounds) of dry food. Use 150 to 200 IU of vitamin E for moist foods. The acetate derivative of vitamin E is chemically stabilized. It retains its vitamin potency longer than unstabilized forms, but it is ineffective as a feed antioxidant.

Several common feedstuffs can contain chemicals that are toxic or that can reduce the value of other ingredients when they are combined. Toxins in oilseeds, including soybean, cottonseed, and rapeseed, can reduce the growth rate of fish if they are not inactivated fully during processing. Manufacturers must be especially diligent in selecting supplies when these ingredients are used. Fresh or dried eggs should never be fed raw, because a component in the egg white will inactivate the biotin (a vitamin). Many raw seafoods contain an enzyme which has been implicated in paralysis and death of fish from thiamine (another vitamin) deficiency. This enzyme is frequently found in fatty fish and mollusk species used for bait. It is less common in fillets prepared for human consumption, but even these are not without risk. A suspect ingredient can be fed for as long as several months before a problem shows up. By that time it may be difficult to trace the cause.

Live Foods and Their Culture

Early attempts to feed fish in captivity depended largely on live foods. Many species of smaller fish, invertebrate animals, and aquatic plants were evaluated repeatedly by trial-and-error methods to see which best supported captive fish. The live foods that eventually became most popular were those that could be collected easily and that were relatively nutritious.

Despite advances in fish nutrition, there remains a need for fresh foods, including live organisms. The least costly way to rear some fish is still to stock the fry, or even the adult broodfish, directly into ponds that have been fertilized to encourage growth of food organisms. Although many species of fish can take dry foods as fry, others require foods too small for conventional manufacturing methods, such as the minute, floating aquatic plants and animals collectively termed plankton.

Plankton can be netted or trapped from ponds for feeding to fish in tanks. A simple but effective plankton trap is made of a small submersible pump, about 1 meter or yard of flexible hose, a small reflector lamp, and a fine-meshed plankton net. The lamp is suspended just over the surface of the water, often clipped to a small raft, and when turned on at night it serves to attract plankton to the pump suspended below. The plankton is caught when water is pumped into the partially submerged net, which is also suspended from the raft or in a nearby bucket. To keep the animals alive until harvest, the water in the net should be aerated during warm weather.

Some types of live foods, including copepods and *Tubifex* worms, are known to carry fish diseases. The risk is greater for live foods collected from the wild than for those reared in captivity. Some of the animals collected as food also may compete with or prey upon aquarium fish. Marine plankton collected for food should be rinsed with fresh water before feeding, so as to limit disease introductions to a marine aquarium. However, rinses cannot be expected to remove internal parasites from an organism.

In marine fish hatcheries, a rotifer, *Branchionus,* is commonly offered as food for very small larvae. If they are gradually adjusted to reduced salinities, marine *Branchionus* can be used in freshwater aquaria as well. Rotifers need to be replaced more often in freshwater tanks because they perish more quickly than in a marine tank. There are freshwater rotifers, including species of *Branchionus,* but they are cultured less commonly for hatchery use. As the fish grow in size (or for hatchlings of larger size), copepods, small crustaceans or worms, and larvae of certain insects are useful.

Rotifers are stocked at a rate of about twenty organisms per milliliter (0.061 cubic inch) of tank water. When even smaller dinoflagellates, mollusk larvae, or ciliates are used for food, higher concentrations are appropriate. Lower concentrations can be used with larger fry fed larger zooplankton. The concentration can be calculated by drawing a sample of tank water into a calibrated pipette. Count the number of planktonic animals in a one-unit area (for example 0.1 milliliter) in the pipette. Multiply that count by the number of units in one milliliter (0.061 cubic inch) to determine the count per milliliter. Add more food as the count declines due to larval feeding or live-food mortality.

Controlled rearing of many aquatic food organisms is technically complex. Nevertheless, advanced aquarists may want to experiment with rearing microalgae and rotifers or other live foods. Seed cultures can be collected from the wild or purchased from scientific supply companies. Suppliers can be located through advertisements in aquarium magazines or a biology teacher may be able to recommend a source. Instructions for culture can usually be obtained from the same suppliers. Methods for hatching or culturing some of the more common live-food organisms follow.

Infusoria: Infusoria (ciliated protozoans, including *Paramecium*) are among the smallest live foods commonly cultivated for fish larvae. Freshwater and marine ciliates can be isolated and cultivated using the same techniques, with appropriate salinity. Ciliates commonly appear as a result of overfeeding. They reproduce rapidly, causing the water to turn cloudy, and eventually reach concentrations of hundreds or thousands of organisms per milliliter of water. Ciliate cultures can be started from pond or aquarium water and sediment, or they can be bought by mail. Larger organisms can be separated from ciliate cultures by pouring the culture through a fine sieve such as a brine-shrimp net. The choice of tank is not critical; a 1- or 2-liter (or quart) beverage bottle is convenient for experimentation. Suitable foods for culturing ciliates include boiled grain, lettuce, hay, and even dry fish food or dried milk. Commercial "infusoria pellets" also are available. It is important to limit the amount of food, however, or the culture will spoil quickly. When dry foods are used, 0.1 to 0.2 grams of food per liter of water per day is enough. Aerate the culture and keep it away from bright lights to keep out filamentous algae. Transfer about 5 percent of the volume to new media after about a week to start a new culture. Ciliates are very small, so a magnifying glass or microscope is useful to check the cultures. If you remove the airstone from a ciliate (or rotifer) tank, the ciliates will rise to the surface, where they can be dipped out for transfer to fish tanks. Ciliates can also be concentrated with a very fine sieve or a plankton net (20-micrometer mesh).

Microworms: One of the simplest live foods to culture for larval fish is the nematode *Panagrellus*. Also known as microworms, these are suitable as a first or second food for many fish too small to feed on brine shrimp. An excellent growth medium is made from 50 grams (1.5 ounces) of finely ground corn flour and 100 milliliters (0.42 cups) of water. If you add 0.5 milliliters of liquid propionic acid to the medium as a preservative, the culture will not spoil as quickly. Spread the medium evenly on the bottom of the culture vessel to a depth of about 5 millimeters (3/8 inch). Inexpensive clear plastic boxes (sold as shoe boxes) work well. Then sprinkle 0.5 grams (0.015 ounces) or more of live dry bakers' yeast on the surface. Transfer and distribute 5 to 10 percent of an established culture as an inoculum. Place a lid on the box and set it aside in a shaded area at room temperature. After about a week, the worms will begin to swarm up the walls of the box. A new culture should be started when the box is first opened, since the culture is likely to be contaminated during harvest. Worms can be scraped or swabbed from the walls for a week or more. Another method for harvest is to place upright pieces of plastic pipe or small plastic balls (about 5 centimeters or 2 inches in diameter) in the medium. After a few days, when these are covered by worms, they can be removed and the worms rinsed into the fish tank. Microworms are not easily separated from their media by sieves, washing, or sedimentation, since they are nearly the same size and density as their food.

Brine shrimp: Brine shrimp (*Artemia salina*) are minute crustaceans collected from saltwater ponds around the world. They are essential for operating hatcheries for several species. Their unique cysts remain viable when dried, so they can be stored on the shelf to be hatched months later when they are needed. One gram (0.03 ounces) of brine shrimp contains about 250,000 cysts. Each cyst contains a single animal, but not all will hatch. The different strains of *Artemia* vary in size and in nutrient content. Generally, strains from coastal waters are preferred for marine fish hatcheries. Freshwater culturists commonly use less costly strains from inland salt lakes.

Cysts are hatched in more dilute brine (1 to 3 percent noniodized table salt) or in seawater. Some strains hatch best in water of a particular salinity or pH; one should follow the recommendations of the supplier in this regard. One liter

(1.06 quarts) of brine is enough to hatch up to about 10 grams (0.3 ounces) of cysts. An excellent hatching tank can be rigged from an inverted plastic bottle with a conical neck, such as a soft-drink bottle. The bottom should be cut off and the top capped. Keep the tank warm (about 30 degrees C or 86 degrees F) and aerate it to prevent the cysts from settling out. Hatching begins in about twenty-four hours and continues over the next day or so. If the airstone is removed, the cysts will float or will settle to the bottom while the tiny crustacean larvae called nauplii will swim toward a light. Nauplii can be collected by siphoning them into another container. Pour the water through a very fine net to separate the nauplii for feeding to fish. The water can be returned to the hatching tank until the remaining shrimp have hatched. A variety of commercially available hatching devices facilitate separation of the newly hatched shrimp from their empty capsules. Unhatched cysts and empty capsules are said to cause intestinal blockage if they are eaten, and they should not be given to fish with the hatched nauplii. Nauplii hatched from regular (capsulated) cysts should be rinsed in fresh water before feeding them to marine fish in order to destroy diseases and competing organisms that might be transferred to the tank with the food.

Another approach is to decapsulate or remove the capsule from the cysts before hatching. Decapsulation makes separation of nauplii and unhatched cysts unnecessary and also disinfects the cysts. Since it involves the use of strong chlorine bleach, it should be done in a well-ventilated area and where a spill will not cause damage. The procedure is as follows. Add 5 grams (0.15 ounces) of cysts (or less) to 100 milliliters (0.42 cups) of cool tap water in a glass or plastic jar. Use fewer cysts per milliliter of water when larger batches are being decapsulated. Allow them to absorb water for a full hour and then add 75 to 100 milliliters (about 1/3 cup) of liquid bleach (5.25 percent sodium hypochlorite) to the jar. Cool the water with ice to keep the temperature below a lethal 40 degrees C (104 degrees F) when larger batches are treated. Aerate the solution or stir continuously to keep the cysts suspended until they turn from a brown color to orange (about five to ten minutes). Drain through a fine-meshed net and rinse with tap water. Rinse the cysts a second time with a 10 percent vinegar solution (equal to 0.5 percent acetic acid), and follow this with an-

other tap-water rinse. Decapsulated cysts can be hatched immediately or they can be stored in saturated brine (about 30 percent plain table salt) until needed. Cysts stored in brine will remain viable for months at room temperatures, even longer in the freezer.

Artemia begin to lose nutritional value soon after hatching, and they will perish within a few days unless they are fed microscopic foods such as unicellular algae or live yeasts. For these reasons, new batches should be started at daily intervals rather than hatching larger batches less frequently. They will not survive for long in fresh water either, so *Artemia* should be kept in salt water until ready to feed.

Earthworms and white worms: Common red earthworms can be cultured as food for larger fish. Earthworm beds should be located in a cool shaded area protected from freezing or flooding. If confined to a poorly drained or poorly ventilated box, they may suffocate if they don't escape first. Worms rarely leave a healthy bed, but when they do nothing will stop them, so indoor production beds are not recommended. A fenced and covered garden bed, or the corner of a garage or shed is better. Worm beds can be separated from walkways with upright 2-by-10-inch (5-by-25-centimeter) wooden boards. Start the bed fairly small and expand the boundaries as the population grows. Ten thousand or more worms can be grown in the space of a square meter or square yard. An excellent medium for earthworms can be blended from two parts of worm or cricket food (available from larger feed stores) and one part each of coarse wheat bran and dry sphagnum peat moss. Finely ground chicken food may substitute for worm food if the latter is not available. Propionic acid can be added to the medium at a 0.5 percent level to control molds that can grow in spoiling feed. Dry storage and good feeding practices will accomplish the same end.

Start the first bed with a 2- to 3-centimeter (about 1 inch) layer of loose crumbly soil or compost. Sprinkle a thin layer of medium over the surface and spray with water. Then add the worms to the bed. Feed them daily, giving only enough medium to last until the next day and spreading food evenly over the entire bed. The bed should be watered gently every day, taking care to saturate the new food, or the worms will be unable to feed. If the bed is kept too dry, mites eventually will take it over. Flies become a problem only when you overfeed or feed table scraps. Ants or rodents should be controlled by traps or poison baits placed outside of the bed. Adult worms congregate near the soil surface as long as they are fed daily. Tan or brown egg capsules about the size of a typed "o," called worm spawn, will also be found once the worms are breeding. It takes another three months, more or less, for the first hatchlings to reach maturity. Commercial worm farmers usually pick all the worms from a bed after a few weeks of breeding and use the top layer to start a new bed, but the bed will support good populations indefinitely if it is regularly harvested and if good medium is used.

The white worm (*Enchytraeus albidus*) is another popular oligochaete, about 1 centimeter (3/8 inch) long. It is often found in association with red earthworms and can be reared under similar conditions in a separate bed.

Feeding Practices

Proper feeding of aquarium fish requires patience and consideration. It is important to understand the fish and to appreciate differences among different species and at different stages in their life cycles. The amount of food required depends upon the type of food, culture conditions, and individual fish. In the wild, large fish may go for days at a time without feeding, whereas newly hatched larvae may feed almost continuously. Fish generally will not overeat, unless they are fed too infrequently. Two feedings a day are best for most fish, more often for newly hatched fry and perhaps less often for very large individuals.

Before feeding, frozen plankton and chopped meats are usually thawed and rinsed in a fine-meshed net to remove dissolved nutrients that might foul the water. However, some aquarists prefer to offer such foods while still frozen, because frozen foods tend to float and to disintegrate slowly as they thaw. Dry foods and minced moist foods should be fed a little at a time over a period of five to ten minutes or until the fish lose interest. Automatic feeding devices are convenient, but they must be checked frequently to be sure that they are still functional and properly adjusted. Blocks of gels or pastes dropped into the tank for feeding should be removed if not finished after ten to fifteen minutes. Fish may fail to feed for reasons other than satiety, and failure to feed may point to serious disease or water quality problems. Sometimes, a change of diet can increase feeding activity markedly.

Most problems with overfeeding result when wasted food spoils in the fish tank. Ammonia and other products of decay degrade water quality and stimulate disease organisms. Overfeeding can cause severe problems even in the large outdoor fish ponds of commercial fish farms. In an aquarium, it is not unusual for undergravel filters to clog from accumulated debris, including uneaten food. Problems associated with an occasional excess can be minimized by regular aquarium care, including cleaning the external filters, redistributing the bottom gravel, and exchanging water. Bottom-feeding scavengers, such as loaches and the common *Plecostomus,* help to clean up freshwater tanks. Several species of fish and invertebrates can be used in saltwater tanks for the same function.

Selected References

Anonymous. 1963. *Composition of Foods, Raw, Processed, Prepared.* Agricultural Handbook No. 8. Washington, D.C., U.S. Department of Agriculture. (A useful guide to compare the nutrient content of human foods that might be fed to fish.)

Anonymous. No date given. *Culturing Algae* and *Carolina Protozoa and Invertebrates Manual.* Carolina Biological Supply Co., 2700 York Road, Burlington, N.C. (Two pamphlets written by a commercial breeder of live foods.)

Anonymous. 1981. *Nutrient Requirements of Coldwater Fishes.* Washington, D.C.: National Academy Press.

Anonymous. 1983. *Nutrient Requirements of Warmwater Fishes and Shellfishes.* Washington, D.C.: National Academy Press. (The preceding two books, part of a continuously updated series of nutrition handbooks published by the U.S. National Academy of Sciences, are intended for use by professional nutritionists. They summarize information needed for formulation of commercial fish foods.)

Moe, M. A. 1982. *The Marine Aquarium Handbook: Beginner to Breeder.* Marathon, Fla.: Norns Publishing Co.

SeaScope, Vol. 1. Summer 1984. Aquarium Systems, Inc., 8141 Tyler Blvd., Menton, Ohio 44060. (A newsletter written by a commercial breeder of live foods and aquarium fish.)

Contributors

JOHN B. GRATZEK received a bachelor of science degree in biology and chemistry at St. Mary's College in Minnesota, where he studied the parasites of muskrats. Pursuing his interests in animal disease, he was awarded the Doctor of Veterinary Medicine degree from the University of Minnesota in 1956 and a Ph.D. in the study of animal virology from the University of Wisconsin in 1961. Dr. Gratzek presently heads the Department of Medical Microbiology in the College of Veterinary Medicine at the University of Georgia. He is past president of the American College of Veterinary Microbiologists and the International Association for Aquatic Animal Medicine, and serves on the aquaculture committee of the American Association of Animal Health.

HOWARD E. EVANS received his Ph.D. degree in Zoology from Cornell University in 1950 and joined the faculty of the New York State Veterinary College. In 1986, Dr. Evans retired as professor and chairman of the Department of Anatomy and professor-at-large of Biological Sciences to become professor emeritus of veterinary and comparative anatomy. His publications include studies of tooth replacement and taste in fishes; reptile and bird anatomy; and anatomy of the dog. He is an associate editor of the *Journal of Morphology*. From 1960 to 1983, Dr. Evans served as Secretary, Vice-President, and President of the World Association of Veterinary Anatomists.

ROBERT E. REINERT did his undergraduate work at Ripon College in Wisconsin, and received his Ph.D. from the University of Michigan. He began his career with the U.S. Fish and Wildlife Service at the Great Lakes Fishery Laboratory in Ann Arbor, Michigan, before becoming Cooperative Unit Leader at the University of Georgia's Cooperative Fish and Wildlife Unit. From 1980 to the present, he has been on the faculty of the School of Forest Resources at the University of Georgia, teaching courses on fish physiology, fish biology, and computer applications. He has written numerous scientific articles, book chapters, and popular articles on fish physiology and aquatic toxicology.

ROBERT A. WINFREE is a research physiologist with the Tunison Laboratory of Fish Nutrition, a branch of the U.S. Fish and Wildlife Service. At his Hagerman, Idaho, laboratory in the Snake River Canyon, he is leading research to improve the health and viability of hatchery-produced fish through applied nutrition. Dr. Winfree lived for several years in Florida, during which time he traveled throughout the Caribbean region working with freshwater and marine aquarium fish, both in research and in commercial production. He has also served as a technical and exhibit consultant for the world's largest totally recirculating marine aquarium system, at Walt Disney's EPCOT Center.

Index of Scientific Names

Subject Index